LECTURE NOTES ON IMPEDANCE SPECTROSC<

Lecture Notes on Impedance Spectroscopy

Measurement, Modeling and Applications

Editor

Olfa Kanoun

Chair for Measurement and Sensor Technology,
Chemnitz University of Technology, Chemnitz, Germany

CRC Press
Taylor & Francis Group
Boca Raton London New York Leiden

CRC Press is an imprint of the
Taylor & Francis Group, an **informa** business

A BALKEMA BOOK

First issued in paperback 2017

CRC Press/Balkema is an imprint of the Taylor & Francis Group, an informa business

© 2011 Taylor & Francis Group, London, UK

Typeset by Vikatan Publishing Solutions (P) Ltd., Chennai, India

Published by: CRC Press/Balkema
P.O. Box 447, 2300 AK Leiden, The Netherlands
e-mail: Pub.NL@taylorandfrancis.com
www.crcpress.com – www.taylorandfrancis.co.uk – www.balkema.nl

ISBN 13: 978-1-138-11506-4 (pbk)
ISBN 13: 978-0-415-68405-7 (hbk)

Lecture Notes on Impedance Spectroscopy – Kanoun (ed)
© 2011 Taylor & Francis Group, London, ISBN 978-0-415-68405-7

Table of contents

Lecture Notes on Impedance Spectroscopy – Kanoun (ed)
© *2011 Taylor & Francis Group, London, ISBN 978-0-415-68405-7*

Preface

Impedance Spectroscopy is a powerful measurement method used in many application fields such as electrochemistry, material science, biology and medicine, semiconductor industry and sensors. Using the complex impedance at various frequencies increases the informational basis that can be gained during a measurement. It helps to separate different effects that contribute to a measurement and, together with advanced mathematical methods, non-accessible quantities can be calculated.

Dealing with Impedance Spectroscopy in general requires competences in several fields of research, such as measurement technology, electrochemistry, modeling, mathematical and physical methods and nonlinear optimization. Depending on the specific challenges of the considered application there are generally more efforts to be done in one or two specific fields. The scientific dialogue between specialists of Impedance Spectroscopy, working with different applications, is therefore particularly profitable and inspiring.

The International Workshop on Impedance Spectroscopy has been launched already in June 2008 with the aim to serve as a platform for specialists and users to share experiences with each other. Since 2009 it became an international workshop addressing more fundamentals and application fields of impedance spectroscopy. The workshop is gaining increasingly more acceptance in the scientific and industrial fields.

This book is the first in the series Lecture Notes on Impedance Spectroscopy. It includes the proceedings of the International Workshop on Impedance Spectroscopy (IWIS'09). The proceedings are a set of presented contributions of world-class manuscripts describing state-of-the-art research in the field of impedance spectroscopy. It reports about new advances and different approaches in dealing with impedance spectroscopy including theory, methods and applications. The book is interesting for research and development in the field of impedance spectroscopy.

I thank all contributors for the interesting contributions, for confidence and for having patience with us during the preparation of the proceedings.

Prof. Dr.-Ing. Olfa Kanoun

Modeling of impedance spectra

Lecture Notes on Impedance Spectroscopy – Kanoun (ed)
© 2011 Taylor & Francis Group, London, ISBN 978-0-415-68405-7

Impedance of nonlinear current or voltage dependent devices

J. Kowal & D.U. Sauer
Electrochemical Energy Conversion and Storage Systems,
Institute for Power Electronics and Electrical Drives (ISEA), RWTH Aachen University, Germany

ABSTRACT: Many real devices such as batteries are nonlinear and their impedance values depend on either current or voltage. However, impedance analysis is based on the assumption of a linear system. If certain conditions such as sufficiently small excitation amplitude to ensure quasi-linearity and measurements at several operating points are fulfilled, also the impedance of a nonlinear system can be interpreted. Equations to calculate the large signal impedance from small signal impedance measurements of current, voltage and mixed dependent devices are derived and analyzed.

Keywords: component, nonlinear, large signal impedance, small signal impedance, negative resistance, negative capacitance, inductive semicircle, inductive behaviour

1 INTRODUCTION

Impedance spectroscopy is a powerful and widely used tool for the parameterization of simulation models. From a single impedance spectrum of a linear device, both the structure of its equivalent circuit and the parameter values can be extracted. For nonlinear devices, the same information can be obtained, but measurements at several operating points are necessary to obtain the complete behavior. Besides the dependency on the operating point, which means the bias current or voltage, the impedance of a nonlinear device also depends on the excitation amplitude. In order to guarantee quasi-linear conditions and thus usable spectra, the amplitude has to be kept small enough (Barsoukov and Macdonald 2005). Furthermore it has to be considered that—in contrast to the linear case—the measured (small signal) impedance of a nonlinear device is not equal to the large signal impedance which is needed for a time domain simulation model.

There are several publications dealing with nonlinear circuits; already in the 1950s, 1960s and 1970s, researchers focused on the mathematical description of polarized electrodes (Macdonald and Brachman 1954; Macdonald 1954; Macdonald 1955), nonlinear capacitors (Macdonald and Brachmann 1955) and nonlinear networks (Popov and Paltov 1963; Chua 1969; Chua and Lin 1975). Those researchers had to face the problem that they could not employ their theory on a grand scale because of the limitations in computing power. The focus of these works was to calculate current and voltage behaviour of a circuit with given nonlinearities, while the focus in this work is to determine the nonlinearities from current and voltage measurements. More recent publications deal with this problem by the usage of impedance spectroscopy, but they typically only determine the small signal impedance in the frequency domain, e.g. (Darowicki 1994; Darowicki 1997; Darowicki 1998).

In a previous paper by the authors, a procedure was developed to determine the large signal impedance of a voltage dependent device from its small signal impedance (Kowal and Sauer 2009). Starting from the differential equation of the current-voltage relationship, the equation is linearized after an AC and DC perturbation. The resulting AC part is transformed into the frequency domain giving differential equations for the small signal impedance elements. In this paper, the method is applied to current and mixed dependency.

The simulations presented in the following show a flipping of semicircles that is caused by the nonlinearities. In practice, similar spectra are measured for different systems. Especially if a semicircle switches from being capacitive to being inductive, this could be interpreted as a change of the equivalent circuit from a parallel connection of resistance and capacitance (capacitive semicircle) to a series connection of resistance and inductance (inductive semicircle). At first sight, this makes sense physically because then only positive equivalent circuit elements are used. However, we will show in this paper that it is not necessary to change the equivalent circuit if negative small signal resistances and capacitances are accepted.

2 IMPEDANCE OF CURRENT AND VOLTAGE DEPENDENT DEVICES

Many electrochemical processes can be described as a current or voltage dependent RC parallel circuit. Fig. 1 shows the equivalent circuit of a nonlinear RC circuit that is considered here. The voltage-dependent large signal capacitance Clsi and resistance R_{lsi} of a RC circuit can be calculated from equations 1 and 2, [12].

$$C_{lsi}(U_{DC}) = \mathrm{Im}\left\{\frac{Y_{ssi}(U_{DC})}{\omega}\right\} = C_{ssi}(U_{DC}) \tag{1}$$

$$R_{lsi}(U_{DC}) = \frac{U_{DC}}{\int_0^{U_{DC}} \mathrm{Re}\{Y_{ssi}(u)\}du + c} = \frac{U_{DC}}{\int_0^{U_{DC}} G_{ssi}(u)du + c} \tag{2}$$

where Y_{ssi} is the measured small signal admittance, $\omega = 2\pi f$ is the frequency and U_{DC} is the bias voltage of the operating point. The small signal capacitance C_{ssi} is used as an abbreviation for $\mathrm{Im}\{Y_{ssi}\}$ and the small signal conductance G_{ssi} is used as an abbreviation for $\mathrm{Re}\{Y_{ssi}\}$. Using the same procedure for a current dependent RC circuit gives similar equations:

$$C_{lsi}(I_{DC}) = \frac{C_{ssi}(I_{DC}) \cdot R_{ssi}(I_{DC})}{R_{lsi}(I_{DC})} \tag{3}$$

$$R_{lsi}(I_{DC}) = \frac{\int_0^{I_{DC}} R_{ssi}(i)di + c}{I_{DC}} \tag{4}$$

where $R_{ssi} = 1/G_{ssi}$ is the small signal resistance of the circuit.

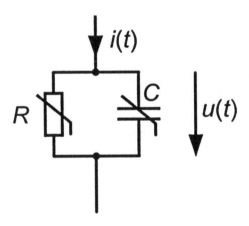

Figure 1.   RC circuit.

4

3 SIMULATION OF CURRENT AND VOLTAGE DEPENDENT DEVICES

Fig. 2 shows the given voltage dependencies of the large signal parameters for the resistor and the capacitance and the resulting simulated impedance spectra of the RC circuit. It can be seen that for higher voltages the real part of the small signal impedance becomes negative although the large signal parameters are always positive in the considered voltage range. This behaviour can be explained by considering the equation to calculate the small signal conductance from the large signal conductance $G_{lsi} = 1/R_{lsi}$, which is the inverse of equation 2:

$$G_{ssi}(U_{DC}) = \frac{dG_{lsi}(u)}{du}\bigg|_{u=U_{DC}} \cdot U_{DC} + G_{lsi}(U_{DC}) \qquad (5)$$

It can be seen that the small signal conductance depends on the derivative of the large signal conductance and if the first term of equation 5 becomes sufficiently negative, the small signal impedance becomes negative. This relationship is illustrated in Fig. 3: For voltages larger than 5V, the slope of the large signal conductance is negative and above about 7V, also the small signal conductance is negative. Fig. 4 shows the simulated impedance spectra (left hand picture) for a given current dependency of R and C (right hand picture). Again both large signal parameters are always positive in the considered current range, but for high

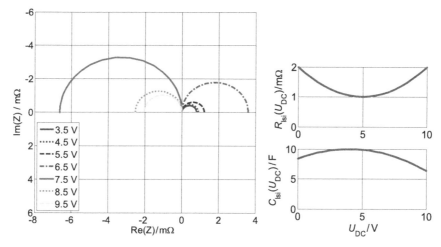

Figure 2. Simulated small signal impedance of an RC element according to Fig. 1 (left) with voltage dependent large signal impedance characteristics for R and C (right).

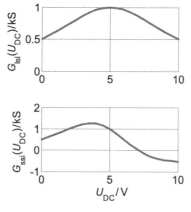

Figure 3. Large signal (upper figure) and small signal (lower figure) conductance corresponding to 2.

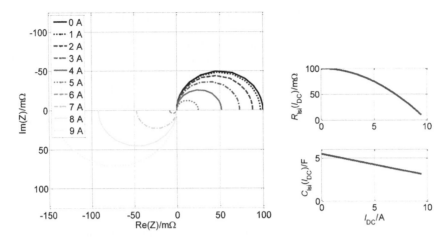

Figure 4. Simulated small signal impedance of an RC element according to Fig. 1 (left hand side) with current dependent large signal impedance characteristics for R and C (right hand side).

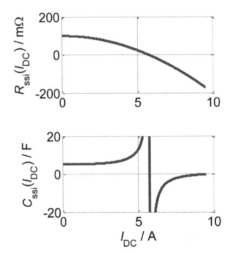

Figure 5. Small signal impedance corresponding to 4.

currents, both the real and the imaginary part become negative. Similarly to voltage dependency, this behaviour can be explained using the relationships between small and large signal impedance (inverse of equations 3 and 4):

$$C_{ssi}(I_{DC}) = \frac{C_{lsi}(I_{DC}) \cdot R_{lsi}(I_{DC})}{R_{ssi}(I_{DC})} \tag{6}$$

$$R_{ssi}(U_{DC}) = \left.\frac{dR_{lsi}(u)}{di}\right|_{i=I_{DC}} \cdot I_{DC} + R_{lsi}(I_{DC}) \tag{7}$$

Equation 7 shows that also the current dependent small signal resistance depends on the derivative of the large signal resistance and if it is sufficiently negative, the small signal resistance becomes negative. Since also the small signal capacitance (equation 6) depends on the small signal resistance, it becomes negative for the same current range. This is illustrated in Fig. 5.

6

4   MIXED DEPENDENCY

Also the equations of mixed dependencies, e.g. $R(u)$ with $C(i)$ and $R(i)$ with $C(u)$, can be derived applying the same procedure. As in the cases of pure current or voltage dependency, the large signal resistance can be calculated independently from the capacitance according to equation 2 for voltage dependent resistance and equation 4 for current dependent resistance. However, the calculation of the large signal capacitance can depend on the large and small signal resistance. The equations for the large signal capacitance for all four cases of current or voltage dependencies are given in Table 1. It can be seen that the capacitance equation depends on the dependency of the resistance. If the resistance is voltage dependent, the large signal capacitance is equal to the small signal capacitance and if the resistance is current dependent, the large signal capacitance depends on the large and small signal resistance as well. The equations were verified by simulated impedance spectroscopy of the four dependencies of RC circuits: A sinusoidal current with a fixed frequency is applied to the parallel connection of resistance and capacitance and the impedance is calculated from the FFT of current and voltage. This is repeated for a set of frequencies for each direct current offset. Fig. 6 shows the Simulink model used for simulation. For each of the four combinations, the large signal resistance is calculated according to equation 2 for voltage depend-

Table 1.   Capacitance equations for all combinations of current and voltage dependent resistance and capacitance.

	$R(u)$	$R(i)$
$C(u)$	$C_{lsi}(u) = C_{ssi}(u)$	$C_{lsi}(u) = C_{ssi}(u) \cdot \dfrac{R_{ssi}(i)}{R_{lsi}(i)}$
$C(i)$	$C_{lsi}(i) = C_{ssi}(i)$	$C_{lsi}(i) = C_{ssi}(i) \cdot \dfrac{R_{ssi}(i)}{R_{lsi}(i)}$

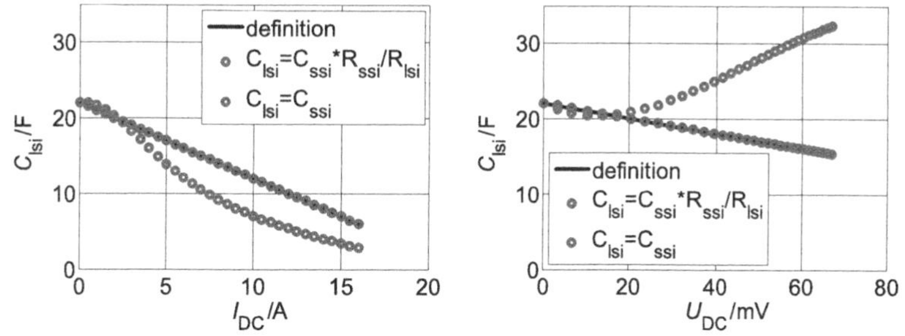

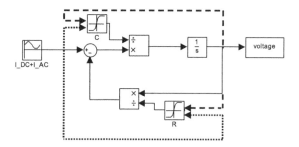

Figure 6.   Comparison of possible equations to calculate the large signal capacitance from simulations of mixed dependent RC circuits. Left hand figure: $R(u)$ with $C(i)$, right hand figure: $R(i)$ with $C(u)$.

Figure 7.   Simulink model for simulated impedance spectroscopy of parallel connection of current and voltage dependent resistance and capacitance.

ent resistance and equation 4 for current dependent resistance. Afterwards, the large signal capacitance is calculated with both equations 1 and 3 and the resulting curve is compared to the given characteristic. Fig. 7 shows the resulting curves for the two possibilities of mixed current and voltage dependency, verifying that the equations in Table 1 are correct.

5 CONCLUSION

Based on the procedure introduced in (Kowal and Sauer 2009) to derive equations for the large signal impedance as a function of the small signal impedance, the equations for an RC parallel circuit with pure voltage, pure current or mixed voltage and current dependency of the elements have been derived and verified by simulations. It turned out that the equation to calculate the large signal capacitance depends on whether the resistance is current or voltage dependent—independent from the kind of dependency of the capacitance. These equations apply to any current or voltage dependent system. Most systems can be modelled both as current and voltage dependent, but typically one dependency is more convenient or makes more sense because of physical or chemical interpretation. According to the decision to model either current or voltage or mixed dependency, the corresponding equations have to be used for parameter extraction from measurement.

Simulations of RC circuits current and voltage dependent elements have shown that for suitable resistance characteristics, the resulting impedance spectra (small signal impedance) show apparently negative resistance semicircles for voltage dependency and apparently negative resistance and capacitance semicircles for current dependency, although both the large signal resistance and capacitance are positive at all times in the performed simulations. This behaviour can be explained mathematically from the corresponding equations for the small signal impedance. The inductive behaviour in the case of current dependent resistance can also be modelled with positive small signal resistance and inductance. However, if such behaviour is observed in measured spectra, it does not necessarily mean that the large signal impedance is also inductive, but a detailed analysis is needed. Before searching for a physical or chemical explanation for inductive behaviour, the large signal impedance should be calculated to find out if it is inductive as well. In the case of voltage dependent resistance, a negative small signal resistance can be observed. Negative large signal resistances are of course impossible for passive circuits, so if such behaviour is measured, it is most probably caused by the nonlinearities of the circuit or otherwise, active elements are present.

REFERENCES

Barsoukov, E. and J.R. Macdonald (2005). *Impedance spectroscopy, theory, experiment and applications* (2nd ed.). Number ISBN: 0-471-64749-7. John Wiley & Sons, Inc.

Chua, L.O. (1969). *Introduction to nonlinear circuit theory*. McGraw Hill.

Chua, L.O. and P.-M. Lin (1975). *Computer aided analysis of electronic circuits: algorithms and computational techniques*. Prentice Hall.

Darowicki, K. (1994). Fundamental-harmonic impedance of first-order electrode reactions. *Electrochim.-Acta. 39*, 2757–2762.

Darowicki, K. (1997). Linearization in impedance measurements. *Electrochimica Acta 42*, 1781–1788.

Darowicki, K. (1998). Differential analysis of impedance data. *Electrochimica Acta 43*, 2281–2285.

Kowal, J., Hente, D. and D.U. Sauer (2009). Model parameterization of nonlinear devices using impedance spectroscopy. *IEEE Transactions on Instrumentation and Measurement 58*, 2343–2350.

Macdonald, J.R. (1954). Static space-charge effects in the diffuse double layer. *J. Chem. Phys. 22*(8), 1317–1322.

Macdonald, J.R. (1955). Note on theories of time-varying space-charge polarization. *J. Chem. Phys. 23*(12), 2308–2309.

Macdonald, J.R. and M.K. Brachman (1954). Exact solution of the debye-hckel equations for a polarized electrode. *J. Chem. Phys. 22*, 1314–1316.

Macdonald, J.R. and M.K. Brachmann (1955). The charging and discharging of nonlinear capacitors. In *Proceedings of the IRE*, Volume 43, pp. 71–78.

Popov, E.P. and I.P. Paltov (1963). Näherungsmethoden zur *Untersuchung nichtlinearer Regelungssysteme*. Akadem. Verlags-Gesellschaft. Geest & Portig K.-G.

Lecture Notes on Impedance Spectroscopy – Kanoun (ed)
© *2011 Taylor & Francis Group, London, ISBN 978-0-415-68405-7*

Time domain simulations of constant phase elements using IIR filter

U. Tröltzsch, P. Büschel & O. Kanoun
Chair for Measurement and Sensor Technology, Chemnitz University of Technology, Chemnitz, Germany

ABSTRACT: Time domain simulations of electrochemical systems like batteries or super capacitors are important for example in electric vehicles or stationary uninteruptable power supplies. Time domain simulations for example are used for predicting the behavior. For time domain simulations adequate online simulation models and methods are required that sufficiently precise describe the underlying physical mechanisms and require only short computation time and small memory. In the frequency domain a lot of well matching impedance models have been developed to describe them. Many of these impedance models models contain constant phase elements that are difficult for direct time domain simulations. In control theory methods are available for the time domain simulation of systems showing constant phase element behavior. In this work these methods are adopted and applied for battery simulation. The transfer function of a constant phase element first is approximated by an ordinary differential equation. Second the battery voltage is simulated using digital filters. Simulation results for a 18650 lithium ion battery during 1000s show a maximum relative voltage deviation of less than 1%. Because of the small resource requirements the method is well suitable for online predictions in battery management systems.

Keywords: constant phase element, battery voltage simulation, impedance spectroscopy, digital filter

1   INTRODUCTION

Time domain simulations of electrochemical systems like batteries or super capacitors are important for example in electric vehicles or stationary uninteruptable power supplies. The energy management systems can use the simulation results to decide about further charging or discharging the battery or the capacitor by not exceeding the end of charge or end of discharge voltage or other limiting factors. Charge or discharge power can be regulated to avoid aging and to extend lifetime. For time domain simulations adequate simulation models are required that sufficiently precise describe the underlying physical mechanisms inside the energy storage. In the frequency domain a lot of well matching impedance models have been developed to describe underlying mechanisms. Many of these impedance models models contain constant phase elements for the description of double layer capacities and diffusion processes (Tröltzsch and Kanoun 2006). Describing the behavior of constant phase elements in the frequency domain can easily be done by an analytical mathematical function. Describing a constant phase element in the time domain is more complex because the inverse fourier or laplace transform will not lead to an ordinary differential equation but to a fractional differential equation. A direct solution of an fractional differential equation in the time domain is not suitable for online simulations in technical systems because it needs a lot of calculation power and memory, especially for systems containing more than one constant phase element. Therefore approximations of constant phase elements are required to use established methods for the solution of ordinary differential equations. These methods requiring only little computational power and memory and are suitable for online simulation. Digital filters, a widely used and well understood technique in communications engineering, are one method for the solution of ordinary differential equations. Digital filters are very easy to implement in microprocessors. The main reason for the development of digital signal processors (DSP's) in the 1980's was the implementation of digital filters. So today very small and powerful DSP's are available that should also be used for online energy storage simulations in mobile and stationary applications to improve their performance and reliability.

This paper discusses the approach to design a digital filter being able to simulate a system showing constant phase element behavior. Therefore first diffusion as an example mechanism is discussed in terms of constant phase elements and fractional differential equations. Second an approximation of the transfer function of a constant phase element using classical transfer function is discussed. Finally a digital filter is designed from the parameters of a classical transfer function. The method is verified experimentally by impedance and pulse pattern measurements. From an impedance spectra parameters of a constant phase element are derived that is used for filter design. The filter is used for simulating the voltage response of the pulse pattern. The quality of the method is assessed by the difference between the measured voltage and the simulated voltage.

2 CONSIDERATIONS ON THE COMPUTATIONAL EFFORT FOR SIMULATING DIFFUSION PHENOMENA USING CONSTANT PHASE ELEMENTS

The temporal and spatial concentration of charge carriers in electrochemical systems is influenced by many mechanisms. One important mechanism is diffusion. Ficks second law is the describing equation for diffusion phenomena and allows the calculation of the charge carrier density c in electrochemical systems. The diffusion constant D is the proportional factor between spatial and time behavior (Hamann and Vielstich 1997).

$$sc = D\frac{\partial^2 c}{\partial x^2} \tag{1}$$

With Ficks first law the relation between charge carrier density c and current density j_{diff} is given. Hereby F is the faraday constant and n is the amount of unit charges of the considered charge carrier species.

$$j_{diff} = nDF\frac{\partial c}{\partial x} \tag{2}$$

For solving Ficks second law for semiinfinite diffusion the following boundary conditions are valid. At the surface of the electrode a constant current density $j_{diff,0}$ is assumed. At an infinite distance from the electrode the concentration of the species is assumed to be constant at c_0.

$$1: j_{diff}(x=0) = j_{diff,0}, \qquad 2: c(x \to \infty) = c_0 \tag{3}$$

For calculating the impedance arising from diffusion the concentration at the electrode surface at $x = 0$ is required.

$$c(x=0) = c_0 + \frac{j_{diff,0}}{nF\sqrt{Ds}} \tag{4}$$

From the concentration at the electrode surface the corresponding diffusion overvoltage η_{diff} can be calculated using the nernst equation. Assuming small voltage changes leads to the linearization of the nernst equation. This linearization is valid if the overvoltages are less than $25mV$.

$$\eta_{diff,0} = \frac{RT}{nF}\ln\frac{c}{c_0} \approx \frac{RT}{c_0 n^2 F^2 \sqrt{Ds}}j_{diff,0} \tag{5}$$

The impedance resulting from diffusion is the ratio of diffusion overvoltage η_{diff} and diffusion current $A \cdot j_{diff}$ at the electrode surface. Hereby A is the area of the electrode.

10

The coefficient σ_W is called Warburg coefficient, the impedance $\underline{Z}_W$ is called warburg impedance Warburg 1899; Muralidharan 1997).

$$\underline{Z}_W(s) = \sqrt{\frac{2\sigma_W^2}{s}} = \sigma_W \frac{\sqrt{2}}{\sqrt{s}} = \sigma_W \frac{1-j}{\sqrt{\omega}} \tag{6}$$

$$\sigma_W = \frac{1}{A} \frac{RT}{c_0 n^2 F^2 \sqrt{2D}} \qquad \left[\frac{\Omega}{\sqrt{s}}\right]$$

Equation 6 represents the warburg impedance in the frequency domain with the frequency $\omega = 2\pi f$ and also in the laplace domain with the complex frequency $s = j\omega$. The laplace representation is important for the constant phase element representation ind the later sanctions. In practice the impedance of diffusion is not observed according to equation 6. Very often the impedance of the diffusion has not a fixed exponent of $\beta = 0.5$ but an exponent of $0 < \beta < 1$. So the warburg impedance will become a constant phase element.

$$\underline{Z}_W(s) = \left(\frac{2\sigma_W^2}{s}\right)^\beta \tag{7}$$

The corresponding impulse response of a system having the transfer function of the constant phase element according to equation 7 can be calculated by inverse laplace transform. The result is the following equation. Hereby Γ is the gamma function.

$$z_W(t) = \left(2\sigma_W^2\right)^\beta \frac{t^{\beta-1}}{\Gamma(\beta)} \tag{8}$$

If the impedance is considered as transfer function in the frequency domain there exists a corresponding differential equation in the time domain. Exciting the system using a certain current i results in a corresponding voltage u. In the frequency or laplace domain the voltage spectra is calculated by multiplying the current spectra with the transfer function.

$$\underline{U}(s) = \underline{Z}_W(s)\underline{I}(s) \tag{9}$$

This multiplication in the laplace domain corresponds to a convolution in the time domain.

$$u(t) = \frac{\left(2\sigma_W^2\right)^\beta}{\Gamma(\beta)} \int_0^t (t-\tau)^{\beta-1} i(\tau) d\tau \tag{10}$$

For time domain simulation of the voltage u of a diffusion element the solution of this integral is required. For short time simulations the solution of this integral is reasonable, for example by approximation the integral by a sum. For long time- and online simulations the computation time and memory usage will rise with simulation time. The solution of this integral is also not reasonable for systems containing more than one diffusion element because they will interact. The interaction leads to more than one convolution integral. To overcome these problems several methods for constant phase element approximation were developed. These methods allow the determination of transfer functions using poles and zeros in the laplace domain in terms of control theory. These transfer functions can easily be simulated using methods for ordinary differential equations or digital filters. The following sections discuss one method to overcome the problem of high computational effort by approximation the constant phase element and to simulate the approximation using digital filters.

3   METHODS FOR CONSTANT PHASE ELEMENT APPROXIMATION

Several methods for approximating the transfer function of a constant phase element by rational fractions are known (Podlubny, Petras, Vinagre, O'Leary, and Dorcak 2002). They mainly base on the determination of poles and zeros of a laplacian transfer function. An elegant method is described in (Oustaloup, Levron, Mathieu, and Nanot 2000). A transfer function corresponding to a constant phase element first is approximated in a certain frequency band. After that the band limited transfer function is again approximated by a polynomial fraction. In (Oustaloup, Levron, Mathieu, and Nanot 2000) the following transfer function is approximated.

$$D(s) = \left(\frac{s}{\omega_u}\right)^n \tag{11}$$

The impulse response of this system can be calculated by the inverse laplace transform. The result is the following equation.

$$d(t) = \frac{1}{\omega_u{}^n \Gamma(-n)} t^{-n-1} \tag{12}$$

The first approximation of the constant phase element is done by the following equation.

$$\left(\frac{s}{\omega_u}\right)^n \approx \left(\frac{\omega_b}{\omega_u} \frac{1 + s/\omega_b}{1 + s/\omega_h}\right)^n \tag{13}$$

The frequency ω_u is defined as center frequency between lower edge frequency ω_b and the upper edge frequency ω_h of the approximation.

$$\omega_u = \sqrt{\omega_b \omega_h} \tag{14}$$

So the transfer function of the constant phase element is replaced by its approximation in a certain frequency band, what is sufficient for many applications.

$$D_A(s) = \left(\frac{\omega_b}{\omega_u} \frac{1 + s/\omega_b}{1 + s/\omega_h}\right)^n \tag{15}$$

The approximation is synthesized from a polynomial fraction containing poles and zeros. If the number of poles and zeros goes to infinity the approximation according to equation 15 will be the same as the infinite product.

$$D_A(s) = \lim_{N \to \infty} D_N(s) \tag{16}$$

$$D_N(s) = \left(\frac{\omega_b}{\omega_u}\right)^n \prod_{k=-N}^{N} \frac{1 + s/\omega'_k}{1 + s/\omega_k} \tag{17}$$

The transfer function according to equation 17 can be used for time domain simulation solving its corresponding ordinary differential equation. The zeros and poles are defined by the corresponding frequencies ω'_k and ω_k that can be calculated according to the following formulas (Oustaloup, Levron, Mathieu, and Nanot 2000).

$$\omega'_k = \omega_b \left(\frac{\omega_h}{\omega_b}\right)^{(k+N+0.5-n/2)/(2N+1)} \tag{18}$$

$$\omega_k = \omega_b \left(\frac{\omega_h}{\omega_b}\right)^{(k+N+0.5+n/2)/(2N+1)} \tag{19}$$

Table 1. Simulation values.

N	=	2	
n	=	−0.594	
ω_b	=	0.0001	$1/s$
ω_h	=	1	$1/s$
ω_u (calculated)	=	0.01	$1/s$

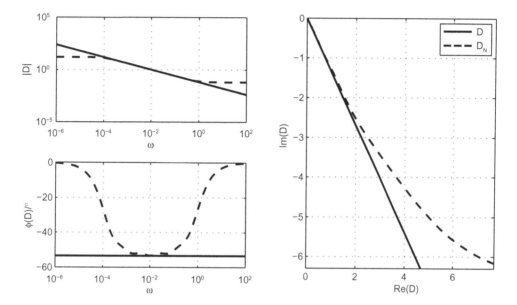

Figure 1. Comparison of the approximation $D_N(s)$ and the constant phase element $D(s)$.

The approximation of the constant phase element now is visualized for some example values. Table 1 shows the values used for simulation. Figure 1 shows the transfer function of the approximation $D_N(s)$ according to equation 17 and the transfer function of the constant phase element $D(s)$ according to equation 11. Within the the approximated frequency range between ω_b and ω_h the absolute value of both transfer function is matching well. Also the angle of the transfer function is in accordance. Outside this frequency range there is a big deviationas expected. Figure 2 shows the deviation between the two approximations $D_A(s)$ according to equation 15 and $D_N(s)$ according to equation 17. For the number of $2N+1=5$ zeros and poles there is an maximum relative deviation of 3.5%. The phase angle maximally differs about 1.4°. So the function $D_N(s)$ according to equation 17 is a very good approximation for the function $D_A(s)$ according to equation 15 also for a small number of poles and zeros.

4 TIME DOMAIN SIMULATIONS

Digital filters are the preferred method for time domain simulations especially for online simulations of technical systems because of their easy implementation using microprocessors. Digital filters require the definition of filter coefficients. These filter coefficients can be determined from poles and zeros of transfer functions in the laplace domain by several transforms. Widely used transforms are the direct transform, the bilinear transform and the

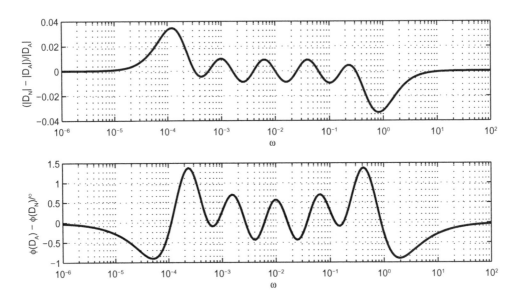

Figure 2. Relative deviation between the approximation $D_A(s)$ and the approximation $D_N(s)$.

impulse invariant transform (van den Enden and Verhoeckx 1990). Because of their very simple realization the direct transform will be used in this article. The poles and zeros of a digital filter in the domain of the z transform are calculated from the poles and zeros in the s domain of the laplace transform. Using the direct transform poles and zeros $z_{z|p}$ in the z domain are calculated using the following equation from poles and zeros $s_{z|p}$ in the s domain. Hereby t_s is the sampling time of the digital system.

$$z_{z|p} = e^{s_{z|p}t_s} \tag{20}$$

The transfer function $D_N(s)$ is the transfer function that should now be used for time domain simulations. For applying the direct transform for digital filter design the transfer function must be rewritten in terms of poles and zeros. So equation 17 will become the following equation.

$$D_N(s) = \left(\frac{\omega_b}{\omega_u}\right)^n \prod_{k=-N}^{N} \frac{\omega_k}{\omega_k'} \prod_{k=-N}^{N} \frac{\omega_k' + s}{\omega_k + s} \tag{21}$$

From this equation the poles and zeros in the laplace domain can easily be obtained.

$$s_{z,k} = -\omega_k' \tag{22}$$

$$s_{p,k} = -\omega_k \tag{23}$$

From the poles and zeros in the laplace domain the poles and zeros in the z domain are calculated using the direct transform according to equation 20.

$$z_{z,k} = e^{-\omega_k' t_s} \tag{24}$$

$$z_{p,k} = e^{-\omega_k t_s} \tag{25}$$

Knowing the zeros and poles in the z domain, the transfer function in the z domain is the following.

14

$$D_Z(z) = k_z \prod_{k=-N}^{N} \frac{z - z_{z,k}}{z - z_{p,k}} \tag{26}$$

$$k_z = \left(\frac{\omega_b}{\omega_u}\right)^n \frac{1}{\left(\prod_{k=-N}^{N} \frac{1-z_{z,k}}{1-z_{p,k}}\right)} \tag{27}$$

The constant k_z determines the DC-amplification of the filter at the frequency $\omega=0$ or $z=1$ respectively. The the DC-amplification of the filter must be the same as the DC-amplification of the approximation $D_N(s)$ according to equation 17. By using equation 27, k_z can be calculated by normalizing the DC amplification values of the laplacian transfer function $D_N(s=0)$ and the transfer function of the digital filter $D_N(z=1)$.

The time domain results of the different systems considered should now be compared using the same parameters used for frequency domain simulation in section 3. Table 1 shows the values used for simulation. The impulse and step responses of the different systems is of interest. The impulse response d of the original constant phase element according to equation 11 is given with equation 12. The impulse response d_N of the approximated system in the laplace domain according to equation 17 can be calculated using state space theory known from control theory. The impulse response d_z of the digital filter according to equation 26 is calculated using its difference equation. Figure 3 shows the impulse responses of the different system considered. All impulse responses in general show the same behavior. The impulse responses of the approximated system decay faster compared to the impulse response of the constant phase element. This is due to their smaller impedance at low frequencies as can be seen in figure 1. The relative deviation between the impulse response of the constant phase element d and the digital filter d_z shows interessting effects. It has its smallest value at approximately $36s$. In the time before the approximation is worse because of the worse approximation at high frequencies and at times after the approximation is worse because of the worse

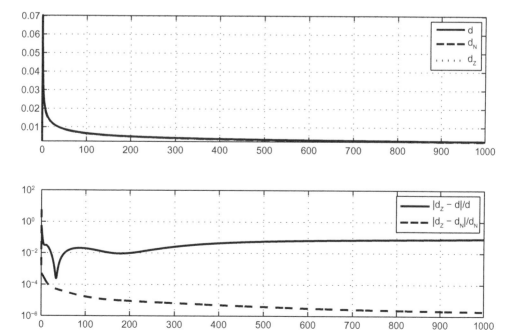

Figure 3. Impulse responses d of the constant phase element, d_N of the system in the laplace domain and d_z of the digital filter and corresponding relative deviations.

15

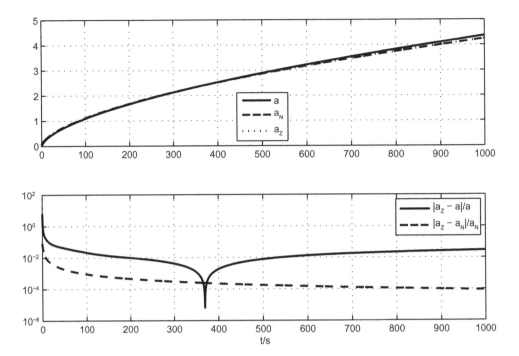

Figure 4. Step responses a of the constant phase element, a_N of the system in the laplace domain and a_z of the digital filter and corresponding relative deviations.

approximation at low frequencies. The relative deviation between the impulse response of the approximated system in the laplace domain according to equation 17 and the digital filter d_z is very small and arises from the direct transform of zeros and poles from laplace domain to z domain. Figure 4 shows the step responses of the different system considered. The step response of the constant phase element a is further rising after $370s$. The step response a_N of the system in the laplace domain according to equation 17 and the step response a_Z of the digital filter tend to reach a stable value. This is required due to system stability.

5 TIME DOMAIN SIMULATION OF BATTERY VOLTAGE

The application of the described method will now be shown by a simulation of the voltage response of a battery. As an example a 18650 Lithium Ion cell with $1.2Ah$ at 80% state of charge was selected.

5.1 Battery model and parameter extraction

A simple battery model according to figure 5 and model equation 28 is used, containing a constant voltage source U_0, a constant electrolyte or series resistance R_E and the constant phase element according to equation 7. First the parameters of the model of the battery are required. The parameters are extracted from an impedance spectrum measured at the battery using nonlinear parameter extraction methods (Kanoun, Tröltzsch, and Traenkler 2006).

$$\underline{Z}_{model}(s) = R_E + \left(\frac{2\sigma_W^2}{s}\right)^\beta \tag{28}$$

16

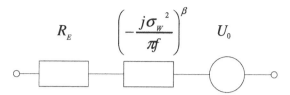

Figure 5. Battery model containing constant phase element used for time domain simulations.

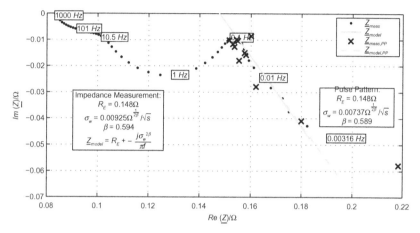

Figure 6. Battery model according to figure 5 fitted to measured impedance data.

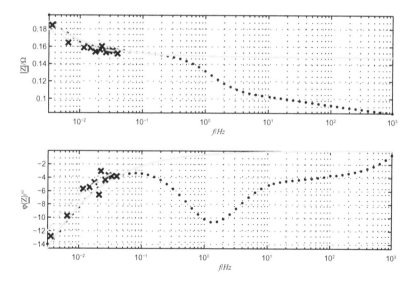

Figure 7. Battery model according to figure 5 fitted to measured impedance data.

Figures 6 and 7 show the Nyquist and Bode diagram of the measured impedance spectrum and the fitted model. The model is matching very well at low frequencies. There are big deviations at high frequencies. Because diffusion occurs at low frequencies and the time domain simulation aims on long simulation times, deviations in the high frequency range are neglected.

For verification reasons also the impedance Z_{PP} extracted from the pulse pattern data measured in the time domain is shown. The impedance Z_{PP} was extracted from time domain

17

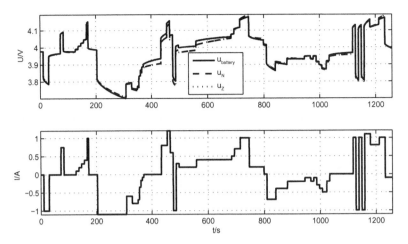

Figure 8.   Time domain simulation.

pulse pattern data applying the fourier transform to the voltage and current data measured at the battery that are shown in figure 8. The model parameters of the model according to equation 28 extracted from pulse pattern data are very similar to the parameters extracted from an impedance spectrum. Also the impedance data show a good agreement.

5.2   *Calculation of filter coefficients*

If the model parameters are known these parameters must be converted for use with the constant phase element representation according to equation 11 instead of constant phase element representation according to the warburg impedance (equation 7). So the battery model will become the following equation.

$$\underline{Z}_{model}(s) = R_E + k\left(\frac{s}{\omega_u}\right)^n \tag{29}$$

Comparing equation 7 and equation 29 gives the relation between the parameters.

$$n = -\beta \tag{30}$$

$$k = \left(\frac{2\sigma_W^2}{\omega_u}\right)^\beta \tag{31}$$

From the known model parameter β the corresponding parameter n of the constant phase element approximation according to equation 17 can be calculated. The frequency range within the approximation is valid is determined by the edge frequencies ω_b and ω_h, which must be selected. The center frequency ω_u is calculated using equation 14. From the known model parameter σ_W and the center frequency ω_u the remaining parameter k can be determined.

5.3   *Voltage response simulation*

The voltage response of the battery can now be simulated. The values given in table 2 are used for simulation.

$$u_N(t) = U_0 + R_E \cdot i(t) + k \cdot \left(i(t) * d_N(t)\right) \tag{32}$$

18

Table 2. Simulation values.

N	=	2	
n	=	−0.594	
ω_b	=	0.0001	$1/s$
ω_h	=	1	$1/s$
ω_u (calculated)	=	0.01	$1/s$
R_E	=	0.14773	Ω
σ_W	=	0.0092473	$\Omega^{\frac{1}{2\beta}}/\sqrt{s}$
k (calculated)	=	0.089214	Ω
U_0	=	3.9776	V

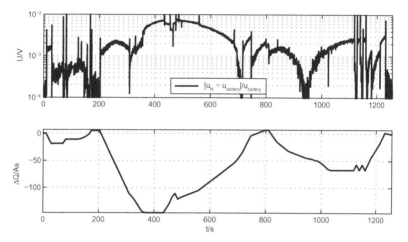

Figure 9. Relative voltage deviation and charge balance.

The voltage response of the system in the laplace domain can be calculated using equation 32. Hereby * is the convolution operator, $dN(t)$ is the impulse response of the approximated system in the laplace domain according to equation 17 and $i(t)$ is the measured time signal of the current.

$$u_Z(t) = U_0 + R_E \cdot i(t) + k \cdot \text{filter}(i(t), D_Z(z)) \qquad (33)$$

The voltage response of the system in the z domain can be calculated using equation 33. The filter operation for filtering the current $i(t)$ using a digital filter having the transfer function $D_Z(z)$ according to equation 26 is done by the filter () function.

Figure 8 shows the simulation results. No significant deviation between the simulation results for u_N and u_Z was observed, the relative deviation was less than 10^{-5}. The relative deviation between the measured battery voltage $u_{battery}$ and the output of the digital filter u_Z is shown in figure 9. Also the amount of charge ΔQ charged or discharged is shown. The relative deviation of the simulated voltage u_Z has its largest values for large changes in charge and for large currents and large current changes. The large deviations at large current changes arise from model properties. The model is not able to simulate fast transients better because in the high frequency range there is a big deviation between model and battery impedance. The deviations at big changes in charge arises from the change of the state of the battery due to the change in state of charge. This change is not considered by changing model parameters. The model parameters are kept constant for the whole simulation. At the end of the simulation the state of charge is the same as in the beginning. Also the relative voltage deviation at the end of the simulation is 10^{-5} as at the beginning of the simulation.

6 CONCLUSION

In this article a method for time domain simulations of battery voltage was presented. The method bases on a constant phase element model for diffusion that is approximated by an ordinary differential equation and simulated using a digital filter. The used model has only 4 model parameters and allows the simulation of battery voltage for at least $1000s$ at a maximum relative voltage deviation of 1%. The computational effort is very small because of the use of digital filters instead of convolution integrals. Further improvements of the presented method can be made by adjusting model parameters with changing battery state for example state of charge, temperature or age. The presented method is well suitable for online prediction of battery voltage in a battery management system. This is for example very important in electrical vehicles for predicting the available power of a battery.

REFERENCES

Hamann, C.H. and W. Vielstich (1997). *Elektrochemie* (3. Auflage ed.). Weinheim: Wiley-VCH.

Kanoun, O., U. Tröltzsch, and H.-R. Traenkler (2006). Benefits of evolutionary strategy in modeling of impedance spectra. *Electrochimica Acta 51*(8–9), 1453–1461.

Muralidharan, V.S. (1997). Warburg impedance-basics revised. *Anti-Corrosion Methods and Materials 44*(1), 26–29.

Oustaloup, A., F. Levron, B. Mathieu, and F. Nanot (2000). Frequency-band complex noninteger differentiator: Characterization and synthesis. *IEEE Transactions on Circuits and Systems I: Fundamental Theory and Applications 47*(1), 25–39.

Podlubny, I., I. Petras, B. Vinagre, P. O'Leary, and L. Dorcak (2002). Analogue realizations of fractionalorder controllers. *Nonlinear Dynamics 29*(1–4), 281–296.

Tröltzsch, U., H.-R. Kanoun, and O. Traenkler (2006). Characterizing aging effects of lithium ion batteries by impedance spectroscopy. *Electrochimica Acta 51*(8–9), 1664–1672.

van den Enden, A. and N. Verhoeckx (1990). *Digitale Signalverarbeitung* (1. Auflage ed.). Braunschweig: Friedr. Vieweg & Sohn Verlagsgesellschaft mbH.

Warburg, E. (1899). Ueber das Verhalten sogenannter unpolarisierbarer Elektroden gegen Wechselstrom. *Annalen der Physik und Chemie 67*(3), 493–499.

Material measurement and testing

Lecture Notes on Impedance Spectroscopy – Kanoun (ed)
© 2011 Taylor & Francis Group, London, ISBN 978-0-415-68405-7

Electrochemical characterization of packed beds by impedance spectroscopy measurements

A. Mahmoud
Universite de Toulouse, Mines Albi, CNRS, Ecole des Mines d'Albi Carmaux, Centre RAPSODEE, 81013 ALBI CT Cedex 09, France

ABSTRACT: The work describes preliminary studies for the application of impedance spectroscopy for the determination of the packed bed conductivity. Electrical conductivity of a packed bed varies with the mobility and affinity of ions with which solid particles are in contact. For very low specific conductivity of the liquid phase, the specific conductivity of the packed bed is enhanced by the presence of solid particles. In contrast, when liquid phase conductivity is high, their contribution to electrical transport becomes more significant.

The paper concerns the investigation of the electrical behaviour of a packed bed, with particular emphasis on bed conductivity determination. For this, different packed-beds have been constituted using potassium chloride solutions, as liquid phase, and synthetic mineral and organic suspensions materials (talc, kaolin, cellulose 101, 105, ...), with different particle size distribution, as solid phase. On the other hand, the case of copper sulphate solutions using Dowex resins of different cross-linking degrees is treated also here.

First, the electrical conductivities of the packed beds were determined, using a model for the conductivity of two-phase media, by measurement of the bed impedance in a lab cell with two facing Pt sheets. Secondly, the conductivity of the bed was analyzed using a "porous-plug" model to understand the current path. Choosing the magnitudes of the components in accordance with the geometry of the dedicated model leads to quantitative agreement between the model theory and the results.

Keywords: electrical conductivity, packed bed, ion-exchange resins, porous-plug

1   INTRODUCTION

A renewed international interest appears during the last decade for electrical techniques based on impedance/or electrical measurement. This type of technique has four main modalities: electrical capacitance technique (ECT), electromagnetic technique (EMT), electrical impedance technique (EIT) and electrical resistance technique (ERT), as shown in Table 1. Each of these techniques has its advantages, disadvantages and limitations. The choice of a particular technique is usually dictated by many, very often contradictory, factors. These include: physical properties of the constituents of multiphase flow, cost of the equipment, its physical dimensions, human resources needed to operate it, and potential hazards to the personnel involved (e.g. radiation). In electrical capacitance technique or ECT (Beck, Byars, Dyakowski, Waterfall, He, Wang, and Yang 1997; Gamio 1997; Plaskowski, Beck, Thorn, and Dyakowski 1995), normally used with mixtures where the continuous phase is non-conducting, the sensor employed is made of a circular array of electrodes distributed around the cross-section to be examined, and the capacitance between all the different electrode-pair combinations is measured. It shows the variation of the dielectric constant (or relative permittivity (ε_r)) inside the sensor area, thus providing an indication of the physical distribution of the various components of the mixture.

Table 1. Comparison of electrical techniques: principles and applications.

Method/Typical arrangement	Measured values	Typical material properties	Typical material
ECT plates	Capacitance C	Permittivity $\varepsilon_r = 10^0 \dots 10^2$ Conductivity $\kappa \prec 0.1$ s/m	Oil, Deionized Water non metallic powders
EIT (ERT) Electrode Array	Impedance (Resistance) Z/R	Permittivity $\varepsilon_r = 10^0 \dots 10^2$ Conductivity $\kappa \approx 0.1 \dots 10^7$ s/m	Water/Saline biological tissue, ion-exchange resins rock/geological materials, semi conductor
EMT Coil array	Self/mutual Inductance L/M	Permeability $\mu_r = 10^0 \dots 10^4$ Conductivity $\kappa \approx 10^2 \dots 10^7$ s/m	Metals, some minerals magnetic materials and ionized water

In principle, ECT has important applications in multiphase flow measurement, particularly gas-oil two-phase flow, which often occurs in many oil wells.

Electromagnetic technique (EMT), based on the measurement of complex mutual inductance, is relatively new and is so far unexploited for process applications. This form of EMT could extract data on permeability (μ_r) and conductivity (κ) distributions. EMT can be selected for processes involving mixtures of ferromagnetic and/or conductive materials. It may applied in following or concentration measurements of ferrite labelled particles in transport and separation processes, foreign-body detection and location for inspection equipment such as used in the food processing, textile and pharmaceutical industries, crack or fault detection, and possibly for water concentration where the water phase is significantly ionized to ensure a high conductivity (Al-Zeibak and Saunders 1993).

In electrical impedance technique (EIT)/electrical resistance technique (ERT) the basic aim is to measure the variations of the electrical conductivity (κ) within an object. Potential applications for EIT/ERT technique are where the continuous phase is electrically conducting while the dispersed phase could be conducting or insulating. Such conditions are typically found in minerals processing applications, medical field, petrochemical field, pressure filtration, mixing, transport and separation processes (Slater, Binley, Versteeg, Cassiani, Birken, and Sandberg 2002; Beck, Williams, and Williams 1995; Brown 2001; Wang, Donvard, Vlaey, and Mann 2000; Bond 1999). Of the four techniques mentioned above, EIT is, in certain cases, the easiest to implement and therefore it was used for the monitoring of ionic state of the packed beds to control operation current density.

2 THEORETICAL MODELS AND ELECTRICAL CONDUCTIVITY

The electrical current flow through packed bed is dependent on the conductivity of the solid and the liquid phases. One of these phases will be the dispersed phase and the other the continuous phases. Several theories have been proposed for the transport of electrical current in mixtures (Helfferich 1962). However, most of the theories are based on the assumptions of either a regular lattice-type solid phase arrangement or a completely random distribution of the components. Neither of these model structures is entirely appropriate for electro-deionisation or filtration cakes. Electrical conductivity of the packed bed varies with the mobility and the affinity of ions with which solid particles are in contact (Helfferich 1962). For very low conductivity of the liquid phase, the specific conductivity of the packed bed is enhanced by the presence of solid particles.

In contrast, when the liquid phase conductivity is high, its contribution to electrical transport becomes more significant.

The Maxwell equation (Maxwell 1873) gives a relationship between the volume fraction of the dispersed phase, ε_s, and the conductivities:

$$\varepsilon_s = \frac{\kappa_b - \kappa}{2\beta\kappa + \beta\kappa_b} \tag{1}$$

where κ is the conductivity of the continuous phase, κ_b is the porous medium conductivity, ε_s is the volume fraction of the dispersed phase of conductivity $\bar{\kappa}$, and β is given by

$$\beta = \frac{\bar{\kappa} - \kappa}{\bar{\kappa} + 2\kappa} = \frac{\alpha - 1}{\alpha + 1} \quad \text{with} \quad \alpha = \frac{\bar{\kappa}}{\kappa} \tag{2}$$

The electrical conductivity of porous medium is complex because of the presence of two conducting phases, namely the dispersed, and the continuous phases, κ.

According to Maxwell, the validity of Equation 1 is limited to small volume fractions. However, various researchers have found that Equation 1 produced good agreement with experimental data over a wide range of void fractions (Neal and Nader 1974; Turner 1976).

If the dispersed phase is assumed to be a non-conducting material, then Equation 9 can be simplified as follows:

$$\varepsilon_s = \frac{1 - K}{1 + 0.5K} \tag{3}$$

where K is the relative conductivity ($K = \kappa_b/\kappa$).

In fact, if Maxwell's equation is expanded as a Taylor series and only the first two terms are considered, the following equation results:

$$K = (1 - \varepsilon_s)^x \tag{4}$$

where the value of the exponent x is 1.5, as originally determined by (Bruggemann 1935). Reference (Rayleigh 1892) attempted to calculate the conductivity of regular arrays of spheres when their interactions could not be neglected. The result of Rayleigh's treatment, after correcting a numerical error in the original paper, is

$$K = 1 + \frac{3\beta\varepsilon_s}{1 - \beta\varepsilon_s - 0.525\left(\frac{\alpha-1}{\alpha+4/3}\right)\beta\varepsilon_s^{10/3}} \tag{5}$$

which is strictly only applicable to when the spheres are arranged in cubic order, and is further limited to moderate values of ε_s.

Another theoretical solution that takes into account the effect of the presence of the dispersed phase on the electrical flux field was proposed by (Meredith and Tobias 1960; Meredith and Tobias 1961) for oil-in-water emulsions:

$$K = \frac{\left(\frac{2+\alpha}{1-\alpha}\right) - 2\varepsilon_s + 0.409\left(\frac{6+3\alpha}{4-3\alpha}\right)\varepsilon_s^{7/3} - 2.133\left(\frac{3-2\alpha}{4+3\alpha}\right)\varepsilon_s^{10/3}}{\left(\frac{2+\alpha}{1-\alpha}\right) - \varepsilon_s + 0.409\left(\frac{6+3\alpha}{4-3\alpha}\right)\varepsilon_s^{7/3} - 0.906\left(\frac{3-2\alpha}{4+3\alpha}\right)\varepsilon_s^{10/3}} \tag{6}$$

For non-conducting particles, Equation 6 can be simplified as following:

$$K = \left(\frac{8(2-\varepsilon_s)(1-\varepsilon_s)}{(4+\varepsilon_s)(4-\varepsilon_s)} \right) \tag{7}$$

However this equation has little advantage over Equation 4 which has an empirically determined value for the exponent x. The relative conductivity with respect to the volume fraction of the dispersed phase has been plotted in Fig. 1, for non-conducting particles, together with theoretical solutions by Maxwell, Rayleigh, Bruggeman and Meredith, respectively (Helfferich 1962; Maxwell 1873; Neal and Nader 1974; Turner 1976; Bruggemann 1935; Rayleigh 1892; Meredith and Tobias 1960; Meredith and Tobias 1961). In contrast to the case of a porous medium of solid inert particles, the geometrical, electrical and chemical properties of particles are sensitive to the chemical composition of the liquid (continuous phase) as adsorption equilibrium prevails. First attempts in modeling often involved regular lattice arrangements, as that suggested by Baron (Helfferich 1962) who considered a statistical cage model. However, most approaches were not valid for solid particles in contact with one another. A more realistic model was developed even earlier (Wyllie and Spiegler 1955; Mahmoud, Muhr, Grvillot, Valentin, and Lapicque 2006). It relies upon the 'porous-plug' model described below. The electrical current is considered to pass through three different paths within the bed (Fig. 2(A)): (i) through alternating layers of particles and interstitial

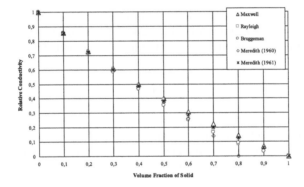

Figure 1.   Relative conductivity versus volume fraction of solid.

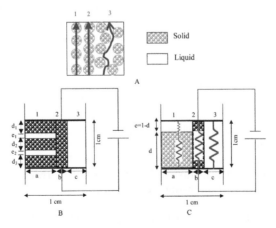

Figure 2.   The "porous-plug" model. A: schematic representation of the three paths which the current can take. B: the simplified model consisting of three conductance elements in parallel. C: Extension of the same model.

26

solution, (ii) through particles in contact with each other, and (iii) in the channel of the liquid phase. In most cases, process (i) predominates, however contribution of (ii) is significant for low-conducting liquids. If the conductivity of the liquid is considerably higher than that of the particles, the third mechanism becomes the most important. The principle can be depicted in terms of electrical resistances in parallel, as shown in Fig. 2(B, C), each of them corresponding to one conduction process.

From the electrical circuit, the backed bed conductivity, κ_b, may be written as the sum of contributions κ_1, κ_2 and κ_3 of the three conductance elements:

$$\kappa_b = \kappa_1 + \kappa_2 + \kappa_3 = \frac{a\kappa\overline{\kappa}}{d\kappa + e\overline{\kappa}} + b\overline{\kappa} + c\kappa \qquad (8)$$

where the parameters a, b and c are the fractional cross-section equivalents of the three elements:

$$a + b + c = 1 \qquad (9)$$

The $(d\kappa + e\overline{\kappa})$ term in Equation 7 expresses the contribution of the liquid, with weight d, and of the solid, weight e in the first conduction process. Parameters d and e obey the relation:

$$d + e = 1 \qquad (10)$$

For a packed bed, the parameters of the model must be estimated to predict the electro-chemical properties of the dispersed phase.

3 MATERIALS AND METHODS

The different materials analyzed in the experiments can be classified into synthetic suspension (mineral and organic) and copper sulphate solutions using Dowex resins. These materials are presented as follows:

3.1 Synthetic suspensions

Five models of synthetic mineral and organic aqueous suspensions with particle sizes ranging from 11–60 μm were used for this study: (talc, kaolin, cellulose105, cellulose101, and microcrystalline cellulose (MMC)). The particle size distribution of these materials was determined using laser diffraction spectrometry (Malvern Master Sizer MS20). Then saturated synthetic suspensions of each material type were made up with a conducting continuous phase, such as diluted potassium chloride solutions.

3.2 Ion-exchange resins (Dowex)

Three ion-exchange resins (Dowex) with cross-linking degree of 2, 4 and 8 DVB% were investigated. Particle sizes ranged from 0.15–0.30 mm for grades 50 WX 2% and 50 WX 4%, and HCR-S 8% particles had a diameter in the range 0.30–0.84 mm. The capacity of resins beds was measured at 0.756, 1.17 and 1.81 eq/l for the 8, 4 and 2% grade, respectively. Resins were tested under the H^+ or Cu^{2+} form: Cu^{2+} form resins were prepared by impregnation of H^+ form sorbents in a column upon continuous percolation by an excess 0.25 M $CuSO_4$ solution. Percolation of the resins with a large volume of 0.25 M sulphuric acid allows their complete conversion to the H^+-form. For both cases, the resins were thoroughly rinsed with deionised water. The conductivity of H^+ resins was measured in dilute sulphuric solutions, while Cu^{2+} resins were investigated in copper sulphate solutions.

3.3 *Experimental set-up*

Conductivity measurements were carried out in a cubic lab cell 2 cm in dimensions, as shown in Fig. 3. Flat platinum electrodes were fixed at two opposite inner walls of the cell. The electrodes were connected to an Autolab PGSTAT 20 potentiostat with a sine function generator. The cell was first filled with the liquid phase using a polyethylene syringe and the velectrochemical impedance was measured at zero cell voltage. The voltage amplitude was 5 mV and the frequency of the signal varied from 30 kHz to 30 Hz, with ten points per decade. The amount of liquid introduced was determined accurately, taking into account the meniscus of the G/L interface, corresponding to an excess of liquid. The error involved in filling the cell was estimated at approx. 2% in the 8 cm³ cell volume. After measurement of the liquid conductivity, the solid was introduced in the cell, and the bed was compacted by vibrating the cell; the excess liquid was removed using a syringe. Compact beds were obtained after the top shiny surface of the liquid over the packed bed disappeared and was replaced by the granular, irregular surface of the packed bed, uncovered by the liquid. Impedance spectra were then recorded. The formerly used solution was thereafter diluted with pure water by 20–30% and the solids were rinsed in the obtained liquid. The experimental procedure of successive measurements with the liquid, then with particles bed, was repeated. Solutions were progressively diluted until their conductivity was of the order of 10e4 S/cm.

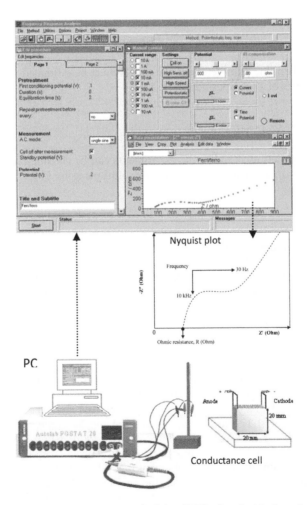

Figure 3. Schematic diagram of measurement principle of EIT of packed bed conductivity.

28

The impedance (Z) is related to the in-phase and out-phase components by:

$$Z(\omega) = Z'(\omega) = jZ''(\omega) \tag{11}$$

where ω is the pulsation, j the complex operator, Z' the resistive component and Z'' the reactive component. The resistance value of the medium investigated can be deduced from the intercept of the curve (Nyquist diagram) or with the real axis. Therefore the intercept of the impedance spectrum recorded at rest potential with the real axis led to the ohmic resistance, R, and the conductivity of the medium, κ, was deduced using the relation:

$$\kappa = \frac{l}{RS} \tag{12}$$

where l is the electrode gap (2 cm), and S the electrode area (2 × 2 cm²). Nevertheless after calibration, the l/S ratio was taken at 0.52 1/cm. Besides, former experiments conducted with packed beds of inert materials with conducting solutions allowed estimation of parameter c. Numerous measurements conducted with the softest model materials led to $c = 0.33$ (0.28 for ion-exchange resins). This value agrees well with the value obtained by (Raghavan and Martin 1995) from modelling of heat transfer in packed beds, with a liquid void fraction of 40%. Then, for these model materials, only two independent parameters had to be determined amongst the series a-e.

4 RESULTS AND DISCUSSION

4.1 *Experimental results*

Fig. 4 and 5 show the conductivity of the different packed beds versus the conductivity of the interstitial solution. The variation depends on the concentration of electrolyte solution and the material grade, as expected. At zero liquid phase conductivity, the current takes its way exclusively through particles which are in contact with on another (the second conductance element (ii)). Even when particle contribution to bed conductivity is very small (because of the small contact area of the particles), the bed conductivity at these conditions can not be equal to zero. The bed conductivity in dilute solutions increases rapidly and is far below than that of the liquid: ions are mainly transferred by the liquid phase. The difference between κ_b and κ decreases for higher concentrations of electrolyte, as shown in Fig. 4 and 5, and the contribution of the liquid phase in the ion transport becomes more significant. At the '*equiconductance*' point $(\kappa_b = \bar{\kappa} = \kappa)$, the bed conductivity is equal to the solution conductivity. This point represents also the conductivity of the corresponding material.

4.2 *Modelling of conductivities using the porous plug model*

Fitting of the experimental curves to the model gave access to the values of the parameters involved in the '*porous-plug*' model. Agreement between experimental data and theoretical variations was generally good, as shown in Fig. 4 and 5. The '*equiconductance*' point, as explained in the previously paragraphs, represents the conductivity of the corresponding material bed and the values obtained are given in Table 2. Table 2 shows good agreement between the theoretical predictions of $\bar{\kappa}$ and the experimental values. Figure 6 schematically illustrates the ion flux within the backed beds, both horizontally and vertically, following the electrical circuit shown in Fig. 2. In Fig. 6, the current flows in the vertical direction. Gray-filled blocks represent the solid phase contribution whereas white blocks correspond to the solution phase. From the left, they illustrate transport through both the solution in contact to the solid phase, the solid phase, and the solution phase, respectively.

For all resins, the current is mainly transported through (i) the alternating layers of particles and solution, and (ii) through particles in contact with each other. Therefore, the ion

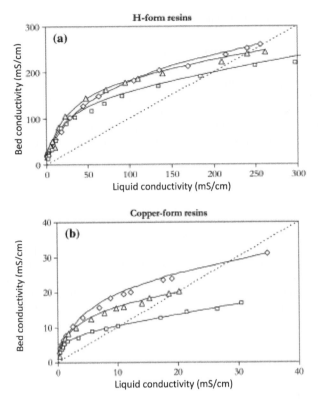

Figure 4. Variation of the bed conductivity with the solution conductivity. (a) H^+-form resins, (b) Cu^{2+}-form resins. Experimental data: (◇) 50 WX 2%, (Δ) 50 WX 4%, (*test*) HCR-S 8%. Solid lines are for are for fitted variations.

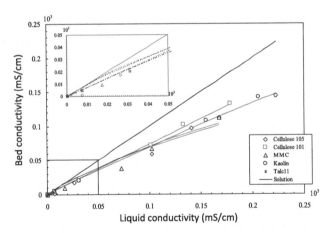

Figure 5. Variation of the bed conductivity with the solution conductivity. Dotted lines are for theoretical results.

exchange resins are more conductive than the solution to be treated, as expected. These resins will significantly reduce the device resistance and increase the available surface area for ion exchange, which is beneficial when performing continuous electro-deionisation operation with dilute solutions. Although the '*porous-plug*' model illustrates the pattern of the ionic flux, the conductivity of the backed bed is an important operating variable for the sake of maximum ion flux.

Table 2. Experimental and theoretical conductivities of solid particles.

Material		Diameter/ Granulometry (μm)	Therotical electrical conductivity $\bar{\kappa}$ (S/m)	Experimental electrical conductivity $\bar{\kappa}$ (S/m)
Dowex 2%	H^+	50 ... 100	26.5	26.6
	Cu^{2+}		2.9	3.1
Dowex 4%	H^+	50 ... 100	26.5	26.6
	Cu^{2+}		2.9	3.1
Dowex HCR-S 8%	H^+	20 ... 50	26.5	26.6
	Cu^{2+}		2.9	3.1
Cellulose 105		20	$0.72 \cdot 10^{-3}$	$0.77 \cdot 10^{-3}$
Cellulose 101		60	$5.65 \cdot 10^{-3}$	$6.05 \cdot 10^{-3}$
Microchrystallin cellulose (MMC)		40	$0.62 \cdot 10^{-3}$	$0.68 \cdot 10^{-3}$
Kaolin		6	$2.10 \cdot 10^{-3}$	$2.17 \cdot 10^{-3}$
Talc		11	$0.47 \cdot 10^{-3}$	$0.45 \cdot 10^{-3}$

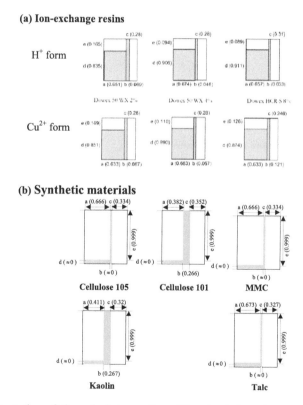

Figure 6. Representation of the ratio between the solid and solution phases; grey blocks are solid phases and white blocks are solution phases. Current flows vertically in the representation.

In case of the synthetic materials, the current is mainly transported through (i) the alternating layers of particles and solution, and (iii) in the channel of the liquid phase. Therefore, the synthetic materials used in this study are less conductive than the solution to be treated, as expected. The conductivity parameters were used to predict the conductivity of synthetic material beds impregnated with a 1 M potassium chloride solutions. Conductivity of the solution was 0.112 S/cm. For the ion exchange resins, the solutions were 1.58 mol/m³ copper sulphate media (i.e. 100 ppm Cu^{2+} species, acidified with sulphuric acid and with a pH near 3.2). Conductivity of the solutions was 1.08 mS/cm.

As shown in Table 2, depending on the degree of cross-linking and the type of the counterion used the conductivity of the solid ($\bar{\kappa}$) decrease by (i) increasing the degree of cross-linking for the resin material and (ii) increasing the valence of the counter ions.

In other words, the degree of cross-linking determines the mesh width and thus the swelling ability of the resin and the mobility of the counter ions in the resin. The latter, in turn, determine the rates of ion exchange and the electric conductivity of the resin. The effects of swelling on the bed conductivity are obvious. Decreasing the amount of divinylbenzene (DVB) in the ion-exchange resin decreases the cross-linking and makes the polymer network coarser. The lower the content of divinylbenzene in the resin, the higher on conductivity.

On the other hand, it can be determined the fraction of the current passing and the results produced by the passage of the current through each of the three elements. In general, the fraction f_e of the current passing through an element is equal to

$$f_e = \frac{\kappa_e}{\kappa_b} \tag{13}$$

Where κ_e and κ_e are the specific conductance of the element and the total plug respectively. Hence, the fraction of current carried by element (i) equals:

$$f_1 = \frac{\kappa_1}{\kappa_b} = \frac{\frac{a\kappa\bar{\kappa}}{d\kappa + e\bar{\kappa}}}{\kappa_b} \tag{14}$$

(a) Ion-exchange resins

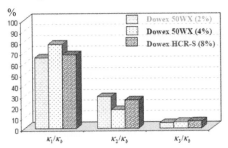

(b) Synthetic materials

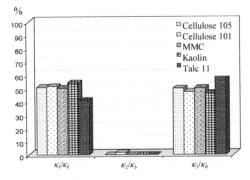

Figure 7.   Representation of the fraction of current carried by elements (i), (ii) and (iii).

Therefore when 1 F passes through the whole plug, only f_1 F coulomb passes through element (i). The fraction of current carried by elements (ii) and (iii) is:

$$f_2 = \frac{\kappa_2}{\kappa_b} = \frac{b\overline{\kappa}}{\kappa_b}; \quad f_3 = \frac{\kappa_3}{\kappa_b} = \frac{c\overline{\kappa}}{\kappa_b} \tag{15}$$

As shown f_1, f_2 and f_3 are given in terms of geometrical parameters determined from conductance measurements. Fig. 7 represents the fraction of the current passing through the elements of the backed bed.

As can be seen from Fig. 7, it is clear that more than 20% flows through the resin phase and more than 70% of the current flows through both the solution and the resins phase alternately based on the porous plug model. In case of synthetic materials, more current is expected to flow through the first and the third paths.

5 CONCLUSION

Electrical impedance spectroscopy is a technique able to identify solid/liquid distribution in packed beds by identification of the bed conductivity and its subsequent modeling as a function of electrical properties of each phase. Electrical conductivity of material in electrolyte solutions was shown to obey the two-phase plug flow model developed decades ago by Wyllie. Moreover, the porous plug model considers not only low concentration but also high concentration solution which has higher conductivity than the ion exchange resin. All the impedance measurements were carried out without flow in the system. Future work will focus on the application of the electrical impedance spectroscopy for the monitoring of ionic state of the bed in continuous electro-deionisation cell. Furthermore, an equivalent circuit need to be de investigated.

REFERENCES

Al-Zeibak, S. and N.H. Saunders (1993). A feasibility study of in vivo electromagnetic imaging. *Phys. Med. Biol. 38*, 151–160.

Beck, M.S., M. Byars, T. Dyakowski, R. Waterfall, R. He, S.M. Wang, and W.Q. Yang (1997). Principles and industrial applications of electrical capacitance tomography. *Measurement and Control 30*, 197–200.

Beck, M.S., A. Williams, and R.A. Williams (1995). *Process Tomography: Principles, Techniques and Applications*. Butterworth-Heinemann.

Bond, J., J.C. Cullivanu, N. Climpson, I. Faulkes, X. Jia, J.A. Knstuch, D. Paylon, M. Wang, S.J. Wang, R.M. West, and R. Williams (1999). Industrial monitoring of hydrocyclone operation using electrical resistance tomography. *Mineral Engineering 12*, 1245–1252.

Brown, B. (2001). Medical impedance tomography and process impedance tomography: A brief review. *Measurement Science and Technology 12*, 991–996.

Bruggemann (1935). Berechnung verschiedener physikalischer konstanten von heterogenen substanzen. i. dielectrizitätkonstanten und leitfähigkeiten der mischkörper aus isotropen substanzen. *Ann. Phys. 24*, 636–679.

Gamio, J.C. (1997). *A High-sensitivity Flexible-excitation Electrical Capacitance Tomography System*. Ph.D. thesis, Institute of Science and Technology, University of Manchester.

Helfferich, F.G. (1962). *Ion Exchange*. McGraw-Hill.

Mahmoud, A., L. Muhr, G. Grvillot, G. Valentin, and F. Lapicque (2006). Ohmic drops in the ion-exchange bed of cationic electrodeionisation cells. *Journal of Applied Electrochemistry 3*, 277–285.

Maxwell, J.C. (1873). *A Treatise on Electricity and Magnetism*. Clarendon Press, Oxford.

Meredith, R.E. and C.W. Tobias (1960). Resistance to potential flow through a cubical array of spheres. *Journal of Applied Physics 31*, 1270–1273.

Meredith, R.E. and C.W. Tobias (1961). Conductivities in emulsions. *Journal Electrochemical Society 108*, 286–290.

Lecture Notes on Impedance Spectroscopy – Kanoun (ed)
© 2011 Taylor & Francis Group, London, ISBN 978-0-415-68405-7

Measurement of the electric conductivity of highly conductive metals with 4-electrode impedance method

R. Gordon, M. Rist, O. Märtens & M. Min
Institute of Electronics, Tallinn University of Technology, Tallinn, Estonia

ABSTRACT: Various metals and their alloys can be characterized, among other measures, also by their electrical conductivity value, for DC and AC. Precise measurements on different frequencies can be used to validate metals, their alloys and their mechanical and physical structure. Specially developed instrumentation (hardware and software) for measuring precisely (with error below 1%) the conductivity of specimens with dimensions of 80 mm*80 mm*3 mm is described. The measurement tolerances are estimated by simulations (in COMSOL Multiphysics environment). Also practical measurements with different materials have been carried out in the conductivity range of about 2.5 ... 60 MS/m and frequency range from DC to 100 kHz AC.

Keywords: conductivity, computer simulation, 4-electrode method, impedance, finite element method, measurement equipment, high current, skin-effect

1 INTRODUCTION

One important parameter of a metal (or alloy) material is it's electrical conductivity (Rossiter 1991). Validation of materials is one important application of electrical conductivity measurements. By measuring on various frequencies, physical and structural non-homogeneity of the material under test can be studied. In current study, one objective has been to design a measurement probe and instrumentation for conductivity measurements of high conductivity metals with a very high accuracy (less than 1% error). The size of the samples is 80 mm*80 mm and the thickness can vary from sample to sample (2–10 mm).

2 ABOUT ELECTRODES AND THEIR POSITIONING

Conductivity measurements can be very sensitive to the quality of electrode contacts—especially when the conductivity of the measured object is higher than the conductivities of the electrodes, contacts and measurement leads. For this task we decided to use 4-electrode method to lessen the errors added by the conductivity and quality of the electrode-probes, leads and contacts. Still the structure of the chosen probes proved to be inadequate to form a constant and reliable contact for the current feeding electrodes. Therefore the use of probes was modified by attaching the cable to the tip area instead of the bottom of the probe thus avoiding inner springs and sliding contacts (Figure 2, shown with ref wire).

Some aspects of 4-point conductivity measurements of metals has been described in works (Bowler and Huang 2005a; Bowler and Huang 2005b; Rietveld, Koijmans, Hall, Harmon, Warnecke, and Schumacher 2003; Koon, Bahl, and Duncan 1989) and general questions of any materials in a book (Schroder 2006). We use convenient electrode positioning, where the 4 electrodes are aligned on the diagonal, symmetrically to the center with 22.6 mm separation (see Figure 1). The outer electrodes are used to feed high current (1–10 A) and the 2 middle electrodes are used for voltage pickup. Conical-tip needle electrodes are used that will be pushed against the sample plate with a force

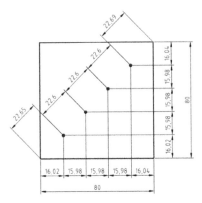

Figure 1. Positioning of the electrodes on the metal sample.

Tip Style 01 (30° tip angle)

Figure 2. All 4 electrodes are conical tip needle electrodes with spring and 30 deg tip angle. A cable is attached to the tip area of the probe. Typical wobble of needle tip is shown on the right.

of 2 N (Ingun test probes—GKS-100, www.ingun.de). This results in a very small contact area for electrodes. At the same time very high current is needed to accurately measure a well-conducting object such as a metal sample with 4-electrode method.

There are two questions that arise from this 4-electrode method:

- Is the positioning accuracy of the electrodes enough to facilitate a 1% measurement accuracy?
- Is the energy density in the electrode tips low enough to avoid heating with the very high current that is needed?

We now answer those questions with computer-simulations of the measurement scenario. The first simulations are conducted using the electrode configuration described above. The metal sample we test in this simulation has the dimensions of 80 mm*80 mm*10 mm and conductivity of 60 MS/m. The maximum wobble described for the probes by the manufacturer (Ingun) is 0.1 mm (see Figure 2). Results from accurate and symmetrical electrode placement is compared with a situation where one current feed electrode is misplaced by 0.1 mm. The modeling and numerical calculation was carried out in Comsol Multiphysics software with Finite Element Method (FEM). The domain (metal sample with electrode contacts) was discretized into 2.4 million elements and the FEM calculation was solved for 3.3 million degrees of freedom (DOF). The element scaling ratio used in the mesh is very high. The element volume in the model varies 2e8 (200 million) times. We use the smallest elements in electrode contact areas to increase calculation accuracy where it is needed the most. At the same time we allow larger elements in other parts of the sample to reduce total computation effort.

This measurement configuration with 1 A current feed gives the result of 0.62 µV potential measured between the 2 middle electrodes. The possible error due to misplacement of the current feed electrode becomes 0.035%. This result is satisfactory for the position accuracy but the microvolt-range signal size might introduce challenges for the measurement equipment.

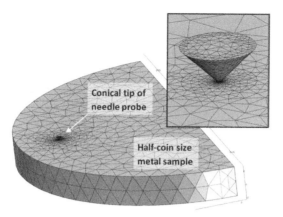

Figure 3. The Finite Element Method mesh of the second simulation with very high element scaling ratio, showing the needle tip in magnified view.

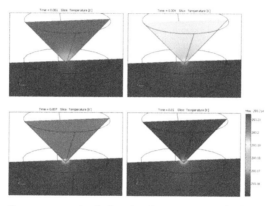

Figure 4. The results of temperature simulation showing temperature in the electrode tip cross-section at 1 ms, 4 ms, 7 ms and 10 ms after the start of 1 A current feed.

We answer the second question with a current density calculation combined with heat-transfer calculation in the conical contact tip. We model the sample in this case as a much smaller object—namely coin-size—and only run calculations on one half (half coin, 1 electrode feeds current, current distribution is assumed symmetrical in the other coin-half). 1 A current is entered through a small cone that has 22 μm contact diameter and led into the sample. The modeling domain (half-coin and conical electrode tip) is discretized into 26000 elements and the coupled electric—and induction currents with heat transfer calculation is performed with 290000 DOF. The element size is also very variable here, being extremely small in electrode contact and largest in edges of the metal sample (see Figure 3). The element volume varies 2e10 (20 billion) times in this case.

The results of this simulation show a current density of 8 GA/m² (8e9 A/m²) in the edge of the contact area. With a measurement that lasts 10 ms, the temperature in the conical electrode tip increases by only 0.64 degrees. Due to very high thermal conductivity of the materials, the energy generated in the very edge of the contact area is immediately dissipated into the contact-probe and the metal sample. Based on the simulation, it is safe to say, that if the current were 10 A and contact area half that of the modeled diameter 22 μm—the thermal conductivity would still dominate over the energy density in the contact region. The current density is highest in the edges of the contact (the 22 μm contact diameter). With higher frequency the energy would be even more concentrated to the edges of the contact, but thermal conductivity would still dominate and would be able to dissipate the heat into the sample and up into the electrode. Therefore we do not expect to find any problems arising from the heating in the contact.

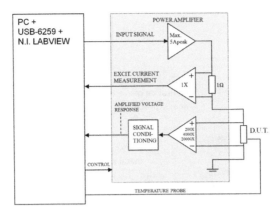

Figure 5. The measurement system schematics. A PC with LabView software on the left, the measurement equipment indicated with a box in the center and the device-under-test on the right.

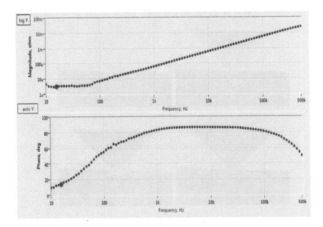

Figure 6. The frequency characteristic of an actual measurement with the developed hardware, just like it is shown real-time with LabView software. The metal sample used in this preliminary test had a high copper-scale conductivity.

3 INSTRUMENTATION: HARDWARE AND SOFTWARE

A special test fixture for holding metal samples for 4-electrode probe has been developed together with hardware and software for high-accuracy conductivity measurements.

The measurement principle used is synchronous detection and is realized in LabView software environment on PC. In order to get the best signal to noise ratio, the excitation waveform is single sinusoid. USB-6259 (multichannel 16 bit, up to 1 Ms/s analog-to-digital and digital-to-analog converters) interface box from National Instruments acts as a bridge between the LabView software and custom electronics.

The developed electronics part consists of a discreet power amplifier, with up to 5 A of output current capability and bandwidth from DC up to 1 MHz and low-noise voltage amplifier, with programmable gain up to 20000. Configuration of the developed interface box is done via USB. In addition, since the impedance of metals is temperature dependant, the temperature of the metal sample is measured using Pt100 temperature sensor. The need for high excitation current and high sensitivity of the voltage pick-up section poses several difficulties in the realization of the measurement system. Sensitive sections can be easily shielded from various capacitive interferences by using Faraday cage. Unfortunately this method is not efficient against strong electromagnetic field, which emanates from the excitation section

38

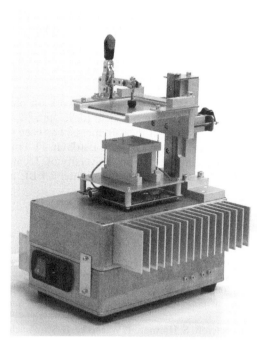

Figure 7. The measurement equipment, complete with special test fixture, 4-electrode probe and a 80 mm*80 mm*3 mm metal sample.

consisting of the power amplifier, distribution wires, probes and the device under test itself. This tends to interfere into the sensitive pick-up amplifier and cause errors that cannot be rejected by synchronous detection as the interfering signal has the same frequency as is used for measurement. To reduce this effect, the area of the frame created by the current loop is kept as small as possible and the fixture with probes and pickup amplifier is placed further from the power amplifier. The current is fed to the sample via a frame and not by cables since this helps to keep the interference more constant making it feasible to be compensated.

The voltage pick-up probes are connected to the high-gain measurement amplifier using custom made twisted pair cable to ensure that the interferences picked up by both leads are equal and rejected by PSRR of the input amplifier.

4   CONCLUSION, RESULTS AND DISCUSSION

The high-accuracy measurement system for measuring electrical conductivity of metal plates has been developed—first the important aspects of measurement have been simulated and then a practical experimental solution has been realized using a 4-electrode measurement probe. Both simulations and experiments show that the accuracy of 1% can be achieved for DC and for AC low frequencies (at 10–100 Hz). At higher frequencies significant challenges come up for simulation as well as the instrumentation (similar, as described in (Hall, Henderson, Ashcroft, Harmon, Warnecke, Schumacher, and Rietveld 2004)). Induced parasitic signals are observed at the test fixture and electronics as well as skin-effect in the sample. We are currently working on the next and considerably improved version of the system, where we expect to overcome or compensate some of the shortcomings in the high frequency capability of the system. It seems necessary to simulate the electro-magnetic fields and interferences that they cause during the development of the next generation measurement device. The software environment chosen for that task is CST STUDIO SUITE.

As an alternative, we are also working on eddy current methods to measure the properties of metals and alloys in wider frequency range (up to 500 kHz). One of our goals is to combine eddy current—and direct contact measurements in one device without a gap in the frequency range between direct contact and eddy current methods.

ACKNOWLEDGEMENT

We would like to thank OÜ Baitaim for the excellent work on physical realization of the measurement test fixture. Current work in this field (development and investigation of signal processing solutions for AC and other measurements) has been supported by Enterprise Estonia (support of Competence Center ELIKO), Estonian IT Foundation (establishment of the sensor-signal processing chair at Tallinn University of Technology), grant ETF7243 and target financing SF0142737s06 and excellence center CEBE at Tallinn University of Technology.

REFERENCES

Bowler, N. and Y. Huang (2005a). Electrical conductivity measurement of metal plates using broadband eddy-current and four-point methods. *Meas. Sci. Technol. 16*, 2193–2200.

Bowler, N. and Y. Huang (2005b). Model-based characterization of homogeneous metal plates by four-point alternating current potential drop measurements. *IEEE Transactions on Magnetics 41*, 2102–2110.

Hall, M., L. Henderson, G. Ashcroft, S. Harmon, P. Warnecke, B. Schumacher, and G. Rietveld (2004). Discrepancies between the dc and ac measurement of low frequency electrical conductivity. In *2004 Conference on Precision Electromagnetic Measurements Digest*.

Koon, D.W., A.A. Bahl, and E.O. Duncan (1989). Measurement of contact placement errors in the van der Pauw technique. *Review of Scientific Instruments 60*, 275–276.

Rietveld, G., L.C.A. Koijmans, Ch. V. and. Henderson, M.J. Hall, S. Harmon, P. Warnecke, and B. Schumacher (2003). Dc conductivity measurements in the van der pauw geometry. *IEEE Transactions on Instrumentation and Measurement 52*, 449–453.

Rossiter, P.L. (1991). *Electrical Resistivity of Metals and Alloys*. Cambridge University Press.

Schroder, D.K. (2006). *Material and Device Characterization*. Wiley-IEEE Press.

Eddy current sensing

Lecture Notes on Impedance Spectroscopy – Kanoun (ed)
© 2011 Taylor & Francis Group, London, ISBN 978-0-415-68405-7

Precise eddy current impedance measurement of metal plates

O. Märtens, R. Gordon, M. Rist & M. Min
Department of Electronics, Tallinn University of Technology, Estonia

A. Pokatilov
AS Metrosert, Tallinn, Estonia

A. Kolyshkins
Riga Technical University, Latvia

ABSTRACT: Properties of conductive materials can be characterized and validated by electrical conductivity. Multi-frequency AC measurements (compared with only DC or single-frequency AC measurements) improve significantly the coverage of validation and give more information about the material properties under investigation. Eddy current measurements, have many benefits, as being nondestructive and faster, compared with 4-contact measurements. Furthermore, for eddy current case results are not directly depending on the exact size of the specimen. Challenge for eddy current measurements is carrying out of absolute measurements, without having reference pieces with known conductivity values. Combining theoretical and experimental studies and development works show, that with single planar coil measurement probe, absolute measurements can be carried out, with accuracies, much better than 10%, for non-magnetic materials with conductivities from 2.5 to 15 MS/m, in the frequency ranges up to 500 kHz. Measurement coil, setup (instrumentation) and software for conductivity measurements (and simultaneously lift-off estimation and compensation) and measurement results are given in the paper.

Keywords: eddy current, planar coil, impedance spectroscopy, conductivity measurement

1 INTRODUCTION

Properties of conductive materials can be characterized and materials validated by measuring of electrical conductivity (Rossiter 1991). Eddy current measurements, using a measurement coil above the metal plate have many benefits, as being nondestructive and faster, compared with 4-contact measurements (Bowler and Huang 2005). Further more, the results (measured material properties) are not directly depending on the exact dimensions and shape of the specimen. Multi-frequency AC measurements (compared with only DC or single-frequency AC measurements) improve significantly the coverage of validation and give more information about the material properties under investigation, as materials properties on the surface and inside can be different (Hall, Henderson, Ashcroft, Harmon, Warnecke, Schumacher, and Rietveld 2004). Challenging task for eddy current measurements is carrying out of so-called "absolute measurements", without having reference pieces with known conductivity values. Furthermore, lift-off compensation can be practical task is real-life setups.

Basis for such approach has been laid by works of Dodd and Deeds, in analytical (Dodd and Deeds 1968) and numerical software (Dodd and Deeds 1975)- last as a BASIC program for PC. Limitation of these works is, that infinite size of the plate has been considered. In approach of (Dodd and Deeds 1975) only active (real) part of the impedance of the coil above the metal plate is calculated. At the same time measuring of the both components of the complex impedance of the coil enables effectively to compensate errors from liftoff uncertainty and so improve significantly the measurement accuracy (Snyder 1995).

So, better starting point for numerical calculations could be results, obtained by Dr. Theodoros P. Theodoulidis and his colleagues (Theodoulidis and Kotouzas 2000), allowing to calculate (simulate) the full (both components of) complex impedance of the coil.

2 MEASUREMENT SETUP AND SOFTWARE

Specimen- material pieces under test, has been used of plate shape, with size 80×80 mm^2 area and 3 mm thick. Industrial LCR meters has been used for measurement of the coil impedance. Different coils have been tested the best results were obtained with the planar (PCB)-coil with N = 80 turns, diameter D = 25 mm (Fig. 1). A software developed by Dr. Theodoros P. Theodoulidis and his colleagues (described also in [7]) has been converted into C/C++ software -by using freeware C/C++ compilers and free libraries for complex calculations, Bessel, Struve and other required functions. Furthermore, an iterative "reverse task" solver has been added into the C/C++ software, making it possible to estimate both actual lift-off and conductivity of the material under test from measured complex impedance of the coil.

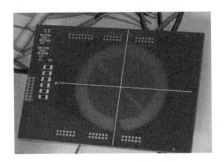

Figure 1.   Planar measurement coil.

Table 1.   Eddy current: conductivity errors.

Nominal conductivity (MS/m)	Relative inaccuracy at frequencies (%)			
	60 kHz	120 kHz	240 kHz	480 kHz
2.157	1.615	1.152	0.668	0.503
9.567	1.168	1.162	0.843	0.942
14.272	−0.614	−0.607	−1.693	−4.291

Table 2.   Eddy current: calculated liftoffs.

Nominal conductivity (MS/m)	Calculated lift-off (mm)			
	60 kHz	120 kHz	240 kHz	480 kHz
2.157	0.179	0.178	0.176	0.175
9.567	0.173	0.175	0.174	0.174
14.272	0.169	0.171	0.170	0.170

3 RESULTS

Experimental investigation shows, that for described setup and software models, accuracy of conductivity measurement better than 10% is achieved, in the frequency range up to 500 kHz. One example set of measurement results is given in the Tables 1 and 2 below.

4 CONCLUSION AND DISCUSSION

Combining theoretical and experimental studies and development and experimental works show, that with single planar coil measurement probe, absolute measurements can be carried out, with accuracies, much better than 10%, for non-magnetic materials with conductivities from 2.5 to 15 MS/m, in the frequency ranges up to 500 kHz. Measurement coil, setup (instrumentation) and software for conductivity measurements (and simultaneously lift-off estimation and compensation) and measurement results has been given in the paper. Further improvement of accuracy is related to more precise using of the additional correction factors- e.g. from finite size of the specimen, parasitic capacitance and other non-idealities of the measurement coil.

ACKNOWLEDGEMENT

Current work in this field (development and investigation of signal processing solutions for AC and other measurements) has been supported by OLAF (the European Anti-Fraud Office), Mint of Finland, Enterprise Estonia (support of Competence Center ELIKO), Estonian IT Foundation (establishment of the sensor-signal processing chair at Tallinn University of Technology), grant ETF7243 and target financing SF0142737s06 and excellence center CEBE at Tallinn University of Technology.

Special thanks to Dr. Theodoros P. Theodoulidis and his colleagues for providing of very valuable software for eddy current calculations (simulations) and Mr. Douglas Reid and Dr. Toomas Kbarsepp for practical suggestions.

REFERENCES

Bowler, N. and Y. Huang (2005). Electrical conductivity measurement of metal plates using broadband eddy-current and four-point methods. *Meas. Sci. Technol. 16*, 2193–2200.

Dodd, C.V. and W.E. Deeds (1968). Analytical solutions to eddy-current probe-coil problems. *Journal of Applied Physics 39*, 2829–2838.

Dodd, C.V. and W.E. Deeds (1975). Calculation of magnetic fields from time-varying currents in the presence of conductors. Technical report, Oak Ridge National Laboratory.

Hall, M., L. Henderson, G. Ashcroft, S. Harmon, P. Warnecke, B. Schumacher, and G. Rietveld (2004). Discrepancies between the dc and ac measurement of low frequency electrical conductivity. In *2004 Conference on Precision Electromagnetic Measurements Digest*.

Rossiter, P.L. (1991). *Electrical Resistivity of Metals and Alloys*. Cambridge University Press.

Snyder, P.J. (1995). Method and apparatus for reducing errors in eddy-current conductivity measurements due to lift-off by interpolating between a plurality of reference conductivity measurements.

Theodoulidis, T.P. and M.K. Kotouzas (2000). Eddy current testing simulation on a personal computer. In *15th World Conference on Nondestructive Testing Roma (Italy)*.

Lecture Notes on Impedance Spectroscopy – Kanoun (ed)
© *2011 Taylor & Francis Group, London, ISBN 978-0-415-68405-7*

High-speed impedance measurement for inline process control of hot-rolled rods

J. Weidenmüller & C. Sehestedt
Laboratory for Sensor Technology and Measurement Engineering, University of Applied Sciences,
Koblenz, Germany

O. Kanoun
Fakultät für Elektro- und Informationstechnik, Technische Universität Chemnitz, Germany

J. Himmel
Laboratory for Sensor Technology and Measurement Engineering, University of Applied Sciences,
Koblenz, Germany

ABSTRACT: For the rod shape measurement of hot rolled round steel bars (rods) the high frequency eddy current method is especially well suited as it requires no contact point and is not limited to bellow the Curie Temperature. Defects of the rod's shape can be detected by measuring the impedance spectrum of the RLC-oscillator. In the first laboratory setup (Weidenmüller, Knopf, Sehestedt, and Himmel 2008) an Agilent Impedance Analyser was used for initial tests. Nevertheless, this setup cannot be applied in a steel plant due to the difficult environmental conditions and low data acquisition rate. Hence, a Vector Network Analyser for passive impedance measurement that is applicable in these surroundings was developed, (Weidenmüller, Sehestedt, Kanoun, and Himmel 2009).

Keywords: Network Analyzer, Surface Testing, Roundness, Roll Displacement

1 INTRODUCTION

In measurement and sensor technology the eddy current principle is one of the most promising methods for noncontact detection of material properties, cracks, distances and pressures (Weidenmüller, Knopf, Sehestedt, and Himmel 2008). Currently, this highly industrial-suited principle is not in use for shape measuring for electroconductive semifinished products, like rods. Three typical defects of ovalities, which occur most frequently during the rod production process, are shown in Figure 1. In order to avoid rejections these typical defects should be detected in the production process as the rolls and the rod's velocity in the production line can be readjusted immediatly. This requires a dynamic measurement system for rod's velocities up to 70 m/s. Currently, in rod shape testing applications measurement systems based on optical sensors are used. Integrating such optical systems into places with harsh environmental conditions, especially close to redhot rods, is difficult. Whereas measurement systems based on high frequency eddy current are much more robust. Also the earlier designed material tracking sensor based on the same high frequency eddy current method confirmed the good industrially applicability of this measurement principle. However, the evaluation electronics implemented in these sensors is not transferable for the rod shape testing application. In order to reconstruct the rod shape, information about the complex impedance spectrum around the resonant frequency of the tuned parallel RLC-oscillator are required. Hence, a robust impedance measurement system was developed as the Agilent Impedance Analyser 4294A used in the laboratory setup is not applicable for tests in a steel plant, Wei08.

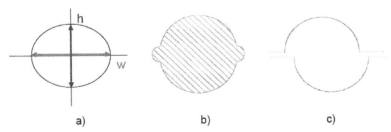

Figure 1. Typical rod defects during production process a) Rod out of roundness b), c) Errors due to roll displacements.

2 PHYSICAL BACKGROUND

The measurement principle can be most easy described using the transformer model (Vyroubal 2004). In this model a primary winding, the detection coil, carries the time-variable current. The time-variable current causes an alternating rotational field, which can be described in terms of the magnetic vector potential $\vec{A}_\Phi$. The magnetic vector potential produces a circular electric field.

$$\vec{E} = -j\omega \vec{A}_\Phi \tag{1}$$

Hence, the induced eddy current on the rod surface is circular as well and has no radial component.

$$i(t)_{eddy} = \int_A \sigma \vec{E} \mathrm{d}\vec{A} \tag{2}$$

As a result, the secondary part of the transformer model, the rod, is characterized with many single turn windings which are fed by respective circular eddy currents. The primary inductance L_c and the secondary inductance L_n are coupled through a shared magnetic field. Consequently, the measured inductance L_σ can be described in terms of the primary inductance L_c and the mutual inductance $M_{c,n}$ caused by the rod.

$$L_\sigma = L_c - (M_{c,1} + \cdots + M_{c,n-1} + M_{c,n}) = L_c + \sum M_{c,n} \tag{3}$$

In this application, the inductance L_σ acting as variable element in a parallel RLC-oscillator and shifts the resonant frequency, as shown in equation 4.

$$\omega_{res} = \sqrt{\frac{1}{L_\sigma C_p}} \sqrt{\sqrt{\frac{2R_c^2 C_p}{L_\sigma} + 1} - \frac{R_c^2 C_p}{L_\sigma}} \tag{4}$$

To gain a better understanding Figure 2 displays the resonance curves for different rod positions inside a detection coil, starting at centre position moving outwards in steps of 1 mm. In this setup the rod diameter is 20 mm and the inner coil diameter is 121 mm. As expected, due to eddy current losses resulting in rising mutual inductance $M_{c,n}$, the effective inductance Ldrops for declining gaps between the coil and the rod (Eq. 3). Due to the interrelationship given by equation 4 the detected resonant frequency increases, (Simonyi 1980; Philippow 2000; Zinke and H. Brunswig 2000).

3 PRELIMINARY CONSIDERATIONS

In the material tracking application, which was successfully tested at the DEW steel plant (Fig. 3), the resonant frequency was detected using a frequency counting system integrated

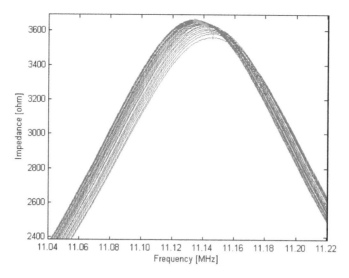

Figure 2.   Resonance curves for different rod positions inside the detection coil.

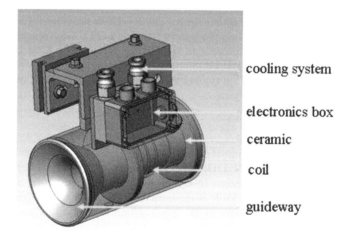

cooling system

electronics box

ceramic

coil

guideway

Figure 3.   SolidWorks © drawing of the material tracking sensor.

into a microcontroller (Himmel, Arend, Otto, and Sehestedt 2006). This method is suitable for material tracking but initial setups indicate that a more extensive data set is required for the rod shape testing application. Therefore a measurement system which also provides information about the complex impedance spectrum near the resonant frequency was developed. For the first experimental setups an Agilent Impedance Analyser 4294A was used (Fig. 2). However, this device is not applicable for such harsh environmental conditions as occurring in steel plants and also not capable of a measurement speed that is sufficient to support process control. Additionally, the advantages such as the wide frequency and impedance range and further analysis options are not required. Consequentially, a new customized HF Vector Network Analyzer (HF VNA) will be implemented into the sensor. The main idea in developing a new HF VNA is to separate the measurement part, including source, bridge and detector, from the control part, that drives the signal source and provides the data collection and processing part. This makes the HF VNA cost efficient and suitable for applications close to hot rolled steel bars as no PC can be located in the immediate vicinity of the production process, (Kiciak 2008a; Kiciak 2008b).

4 HF VECTOR NETWORK ANALYZER

The HF VNA used is based on the measurement of the reflection, the Agilent 4294A on the other hand side utilizes an **RF I-V** method. The **RF I-V** method is suitable for a wide range in impedance as well as in frequency but the reflection measurement has comparable sensitivity in a narrow range around the null balance point that the bridge is matched for. In this chapter the key elements of the designed VNA will be introduced, which are:

- A software controlled synthetized signal source.
- A bridge to separate incident from reflected signals.
- A detector capable of amplitude and phase detection.
- Controls, data collection, and processing.
- The Device Under Test (DUT), the RLC-oscillator.
- OSL calibration standards (Open, Short, Load).

4.1 *Signal source*

As shown in Figure 4 there are two signal sources in this VNA, which are called RF DDS and LO DDS. Both are based on direct digital synthesis (DDS), to be exact two AD9851 are in use. Both DDSs operate at the same frequency, which ranges from nit 50 kHz to 60 MHz with a minimum frequency step size of approximately 0.035 Hz. Hence those DDS appropriate to the rod shape application, as the expected resonance curves of the measuring RLC-oscillator tuned by the rod are well within that range. The RF DDS provides the signal source to the bridge and DUT during reflection measurements. The LO DDS is frequency and phase synchronized with the RF DDS in order to provide the reference signal which is needed later on in the detection part. The RF DDS is always programmed for 0° while the LO DDS is programmed for either 0 or 90° in typical VNA usage. The master oscillator for the DDSs is provided by an 148.34 MHz crystal.

4.2 *Reflection bridge*

The reflection bridge is the terminal for the DUT and necessary to separate the incident (RF DDS IN) from the reflected (RF DDS OUT) signal. There are a lot of hardware options that can be used for a reflection bridge. In this application a bridge with a T1-6T Mini-Circuit transformer was chosen (Fig. 5). It is recommended to place the bridge in a separate box in order to gain best performance of the VNA. Due to the good electrical RF characteristics, repeatability and compatibility with commercial VNA calibration standards, SMA connectors are used. The displayed bridge in figure 5 is designed for the first laboratory setup and will be redesigned for tests in a steel plant, as the VNA must be calibrated before the

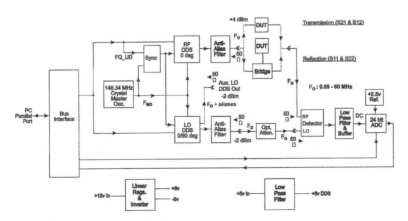

Figure 4.   HF vector network analyser block diagram [8].

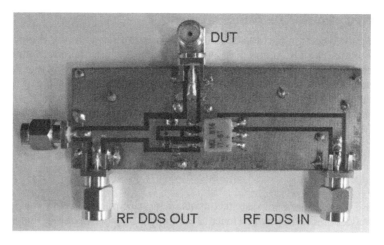

Figure 5. T1-6T reflection bridge.

measurement starts using the OSL-calibration standards. During the production process it is not possible to connect the OSL calibration standard to the bridge manually, hence a software controlled switching system using a microcontroller will be implemented. In this way the VNA can be calibrated via software before the process starts.

4.3 *Detector*

Basically, the VNA detector is realized using a linearized Gilbert cell mixer (MC1496), analog filtering and buffering, a 24 bit analog to digital converter (LTC2440CGN), and a precise and temperature stable 2.5 V reference for the ADC (Fig. 4). Consequentially, the theoretical resolution of the ADC is approximately 0.149 µV. The conversation speed of the LTC2440CGN is adjustable with the fastest performance being 3.5 kHz. As both DDSs are programmed with the same frequency, the voltage at the detector output is DC and contains the desired carrier information. Please consider, this is only obtained by using a synchronised LO DDS. Otherwise if the LO DDS is asynchronous with its RF DDS the information is contained in the detected sidebands. As the phase of the LO DDS can be programmed to values of 0° and 90°, the detector output takes on two DC values that together represent the vector components of the applied RF signal, as shown in equation 5 and 6.

$$V_{DC,0°} = |G_{DET}| \times |V_{RF}| \times \cos(\Phi_{RF} + \Theta_{DET}) + V_{off,0°} \qquad (5)$$

$$V_{DC,90°} = |G_{DET}| \times |V_{RF}| \times \cos(\Phi_{RF} + \Theta_{DET}) + V_{off,90°} \qquad (6)$$

$|G_{DET}|$ is the magnitude of the detector gain
Θ_{DET} is the phase constant related to detector gain
$|V_{RF}|$ is the magnitude of the detector RF input voltage
Φ_{RF} is the phase at the detector RF input
Φ_{LO} is the phase at the detector LO input
V_{off} is the offset voltage with no applied signal at RF input

The DC offset voltages $V_{off,0°}$ and $V_{off,90°}$ are obtained from the 'Open' detector calibration, in this case the magnitude of the voltage at the detector RF input is zero. These values are later on subtracted from all subsequent measurements in the data processing part. With $V_{off,0°}$ and $V_{off,90°}$ known, the constants $|G_{DET}|$ and Θ_{DET} are determined with the 'Short' calibration. Finally, $|V_{RF}|$ and Φ_{RF} are measured when the DUT is connected, (Kiciak 2008a; Kiciak 2008b).

4.4 Data processing

After passing the detection part the desired signals are transferred to the PC via a bus interface. They contain two DC values $V_{DC,0°}$ and $V_{DC,90°}$ that include the information about the reflected voltage $V_{reflection}$. In the PC the calculation mentioned in subchapter 4.3. Detector are performed using the software LabView®. Figure 6 displays the principle of the reflection measurement with a characteristic impedance of the measurement circuit Z_0 of 50 Ω. The Scattering parameter $\underline{S}_{11}$ containing the information about the reflected signal can be calculated using the following equation:

$$\underline{S}_{11} = \frac{\underline{b}_1}{\underline{a}_1} = \frac{V_{reflection}}{V_{towards}} = \Gamma \tag{7}$$

In this case the measurement system consists of just one port, therefore the $\underline{S}_{11}$ parameter equals the reflection coefficient Γ. Hence the complex impedance can be determined with equation 8.

$$Z_x = Z_0 \cdot \frac{\Gamma + 1}{\Gamma - 1} \tag{8}$$

As mentioned above, the reflection measurement has high impedance measurement sensitivity around the null balance point. Thus a 50 Ω bridge is applied, as the expected impedance lies within the high sensitivity range between 10 and 200 Ω (Fig. 7). If the impedance anticipated is out of that range the bridge needs to be matched to the range of interest.

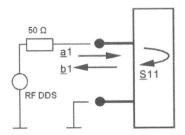

Figure 6. $\underline{S}_{11}$ parameter.

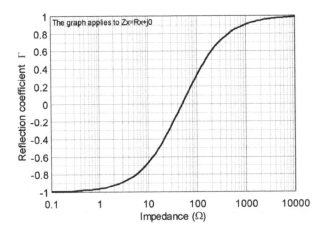

Figure 7.   Relationship of reflection coefficient to impedance for a 50 Ω bridge (Agilent 2009).

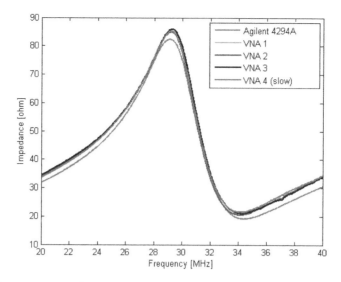

Figure 8. Laboratory test results.

5 RESULTS

Figure 8 displays the resonance curve of the RLC-oscillator measured with different VNAs. The inductance of the detection coil in use is 240 nH and the desired resonant frequency of about 29 MHz is tuned with a variable parallel capacitor. In order to compare both measurement systems this setup is first tested with the Agilent 4294A. After a warm-up time of about 15 minutes of the new designed VNAs, which is recommended in order to gain stable temperature conditions, the four VNAs are tested in the frequency spectrum between 20 and 40 MHz. Whereas the ADC sample rate in one version is set to 7 Hz, the so called slow version, the other three ADCs are set to the fastest performance of 3.5 kHz. As mentioned in a previous chapter, this high speed performance is required in rod shape testing. As a result, the new designed VNAs yield good precision but a static offset of about 3 ohm compared to the Agilent 4294A. Possible reasons therefore are:

- Temperature drift
- The use of different calibration standards
- Unshielded experimental setup

Those possible reasons will be investigated in the next step of development.

6 CONCLUSION

The new high frequency eddy current rod shape testing approach has several advantages compared to currently in use sensor systems. The promising test results using the laboratory setup (Weidenmüller Knopf, Sehestedt, and Himmel 2008) can be verified in a steel plant as the designed HF VNA is applicable into places with harsh environmental conditions. Due to the compact and robust front-end electronic the HF VNA can be located close to the detection unit which is necessary in order to gain stable measurement conditions. Also the data acquisition speed is sufficient to support process control. In the next step of development the frond-end electronic will be placed into a temperature controlled and shielded box. Additionally, the OSL calibration standards will be automatically connected to thereflection bridge via a software controlled switching system. After comparing the accuracy and sensitivity of the designed impedance measurement system with the test results acquired using an Agilent 4294A, the first prototype will be constructed. Finally, referring to the obtained database a mathematical approach has to be developed to reconstruct the rod shape.

ACKNOWLEDGEMENTS

The Foundation Rhineland-Palatinate for Innovation (Stiftung Rheinland-Pfalz für Innovation) is founding this research project (number 794) from 01/2008 up to 12/2011.

REFERENCES

Agilent (2009). Impedance measurement handbook. Website. http://cp.literature.agilent.com/litweb/pdf/5950-3000.pdf.

Himmel, J., S. Arend, A. Otto, and C. Sehestedt (2006). Produktionsflusskontrolle in Walzwerken mit elektromagnetischen Streufeldern. In *Sensoren Signale Systeme*.

Kiciak, P. (2008a). An HF Vector Network Analyzer—Part 1. Website. http://n2pk.com/VNA/n2pk_vna_pt_1_ver_c.pdf.

Kiciak, P. (2008b). An HF Vector Network Analyzer—Part 2. Website. http://n2pk.com/VNA/n2pk_vna_pt_2_ver_c.pdf.

Philippow, E. (2000). *Grundlagen der Elektrotechnik*. Verlag Technik.

Simonyi, K. (1980). *Theoretische Elektrotechnik*. VEB Deutscher Verlag der Wissenschaften.

Vyroubal, D. (2004). Impedance of the eddy-current displacement probe: The transformer model. *IEEE Transactions on Instrumentation and Measurement 53*, 384–391.

Weidenmüller, J., C. Knopf, C. Sehestedt, and J. Himmel (2008). Rod shape testing by high frequency eddy current—the experimental setup. In *I2MTC Conference IEEE IM Chapter*.

Weidenmüller, J., C. Sehestedt O. Kanoun, and J. Himmel (2009). Rod shape testing by high frequency eddy current—passive impedance measurement. In *SSD Conference IEEE IM Chapter*.

Zinke, O. and H.H. Brunswig (2000). *Hochfrequenztechnik 1- Hochfrequenzfilter, Leitungen, Antennen*. Springer Verlag.

Bio and environmental applications

Lecture Notes on Impedance Spectroscopy – Kanoun (ed)
© 2011 Taylor & Francis Group, London, ISBN 978-0-415-68405-7

Analysis of body segments using bioimpedance spectroscopy and finite element method

M. Ulbrich, L. Röthlingshöfer, M. Walter & S. Leonhardt

RWTH Aachen, University of Technology, Chair for Medical Information Technology (MedIT), Germany

ABSTRACT: Elderly people and geriatric patients often suffer from dehydration which can be diagnosed by bioimpedance spectroscopy (BIS). In this work, the capability of segmental BIS to reflect the total body composition is analyzed. Therefore, measurements of body segments have been simulated using the finite integration technique (FIT) and the results have been verified by measurements on subjects. We show that results obtained from segmental BIS measurements correlate with wholebody bioimpedance spectroscopy measurements. Furthermore, simulations of segmental BIS have been successfully accomplished and the simulation results show similar characteristics compared to measurements performed for reference.

Keywords: bioimpedance spectroscopi, finite integration technique

1 INTRODUCTION

Since the beginning of the 20th century, a demographic change is observable in Europe which leads to a steadily aging society. Looking at Germany as an example, the average life expectancy of women is supposed to rise from 74.2 to 87.1 years between 1913 and 2035. Important factors are the continuous progress in medical technology and the decreasing birth rate of actually around 1.4 children per couple (Statistisches Bundesamt 2007). Hence, society will have to face more and more geriatric patients, which leads to additional costs and burdens on medical personnel. Thus, it is reasonable to establish methods to treat elderly people more cost-effectively and more easily by improving diagnostics for certain diseases that prevalently occur among elderly people.

One of these diseases is dehydration, which describes a water deficiency in the body. This deficiency is primarily caused by a lower thirst sensation and a disturbed hormone balance (Lavizzo-Mourey 1987). Also athletes can suffer from dehydration during physical exercise, which can lead to lifethreatening situations. The project NutriWear, funded by the German Federal Ministry of Education and Research (BMBF), aims at developing a wearable system to monitor the nutrition and hydration status based on intelligent textiles. What is more, the patient shall be monitored 24 hours per day (Bundesministerium für Bildung und Forschung 2007). Such a system can be used for preventive monitoring of dehydration.

Dehydration can be easily and cost-effectively diagnosed by bioimpedance spectroscopy (BIS). Currently, BIS is not commonly used as diagnostic method because it is not considered to be valid, even though there are a lot of studies implying the opposite (Wabel 2009). One possible reason is that many processes in the human body during BIS are widely unknown. One way to analyze where the current paths run, and which tissue contributes significantly to the measurement result, is to use computer simulations employing FIT (Clemens 2001). In the following, correlations of segmental BIS and wholebody BIS measurements shall be analyzed and compared with simulations of segmental BIS measurements.

2 BIOIMPEDANCE SPECTROSCOPY

The basic idea of BIS is that each cell consists of a capacitive cell membrane, separating extra-cellular and intracellular space. When injecting an alternating current into a certain tissue, these membranes lead to a current flow between the cells through the extracellular water (ECW) at low frequencies. At high frequencies, the membranes are no barrier for the current so that it flows through all cells now as a function of ECW as well as intracellular water (ICW) (cf. Fig. 1).

There are three major dispersion regions: α (mHz–kHz), β (0.1–100 MHz) and γ (0.1–100 GHz). The frequency range between 1 kHz and 10 MHz within the α dispersion region is generally the most interesting one for diagnosis because physiological and pathophysi-ological processes lead to changes in body impedances with high dynamics in this range. The reason for this is polar proteins and organelles which behave like dipoles in the alternating field (Grimnes and Martinsen 2000). In addition, in this frequency region safety regulations permit higher alternating currents than in the frequency range below 1 kHz. Thus, the degree of meas-urement accuracy is kept at an optimum. BIS is a method for measuring bioimpedances using a frequency spectrum between 5 kHz and 1 MHz with a current between 500 A and 10 mA.

Usually, whole-body BIS measurements are accomplished using one electrode pair to inject the current and one electrode pair to measure the voltage at wrist and ankle so that four electrodes have to be used for one measurement. The resulting impedance curve represents a complex cole-cole curve (cf. Fig. 2, left).

The model which fits very well to reality is represented by the equivalent circuit for this plot shown on the right side of Fig. 2, whose impedance can be described by the following formula. Here C_m is the membrane capacity, R_e the extracellular, R_i the intracellular resistance and $R_\infty = R_e \| R_i$ (Grimnes and Martinsen 2000).

$$Z(\omega) = R_\infty + \frac{R_e - R_\infty}{1 + (j\omega C_m (R_i + R_e))^{\alpha}} \tag{1}$$

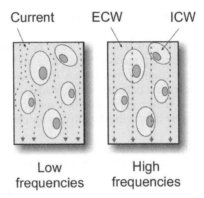

Figure 1.   Current flow in tissue.

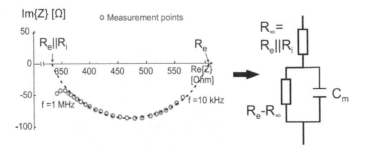

Figure 2.   Whole-body BIS measurement, typical impedance progression and equivalent circuit diagram.

58

Using these resistances, extra- and intracellular volume, total water, fat free mass, fat mass and muscle mass can be calculated (Lorenzo 1997). The capacitor representing the cell membrane is a constant phase element (CPE) because the semi circle of the cole-cole curve has its center somewhere above the x-axis. The parameter α reflects the homogeneity of the tissue and therefore the shift of the semi circle's center (Grimnes and Martinsen 2000).

Classical whole-body BIS analyzes the whole-body impedance between wrist and ankle, assuming the body to be one cylinder with one conductivity for extracellular resistance and one for inter-cellular resistance. Thus, the extra- and intracellular volume can be calculated by the following equations:

$$ECW = \frac{1}{1000} \left(\frac{K_B^2 \rho_e^2}{D_B} \right)^{1/3} \cdot \left(\frac{H^2 \sqrt{W}}{R_e} \right)^{2/3} \quad (2)$$

$$ICW = ECW \left(\left[\frac{\rho_{tbw} \cdot (R_e + R_i)}{\rho_e R_i} \right]^{2/3} - 1 \right) \quad (3)$$

Here KB

is a dimension less shape factor calculated from an atomic measurements,

ρ_e and ρ_{tbw} are the specific resistivities of the extracellular and overall fluid, W is the body weight and H the body height (Matthie 2005).

Approximating the human body as one cylinder is of course an assumption which leads to modeling errors. These errors can be reduced by sub-dividing the body into several segments (Zhu 2006). Segmental and whole-body measurements have been compared to reference methods for detecting the body composition, such as Dual Energy X-Ray Absorpiometry (DEXA), ^{40}K or D_2O solution method.

DEXA technology uses two different x-ray sources for emitting x-rays through the human body. A detector receives the non-absorbed radiation which depends on the density of the irradiated tissue. Thus, bone density and body fat can be calculated. Comparing segmental and whole-body measurements with fat free mass (FFM) assessments by DEXA, both bio-impedance methods generate correct values. In addition, BIS measurement results of single segments show high correlations with results of DEXA analysis of the same segments (Salmi 2003; Bracco 1996).

The second reference method introduced here is the ^{40}K method. ^{40}K is a radioactive potassium isotope which is administered orally. It preferably accumulates in body fat and there emits x-rays when decaying. ECW and ICW measured by segmental BIS and the ^{40}K method show good correlations (Carter 2007).

The other substance taken as reference method is deuterium oxide (D_2O), also called heavy water, which is taken orally and is distributed in all body tissues except fat. By taking a blood sample and measuring the D_2O content in plasma, the total body water can be calculated. Segmental measurements are superior to whole-body measurements when detecting the hydration status compared to D_2O measurements (Thomas 2003).

It has been also shown without reference measurements, that thoracic segmental BIS measurements are better than wholebody measurements concerning the hyperhydration detection for dialysis patients (Nescolarde 2007). To sum it up, the reason why segmental measurements and simulations have been accomplished is that several studies showed that segmental BIS produces reliable and in some cases even better results than conventional whole-body BIS.

3 METHODS

Classically, the body is divided into five segments: arms, legs and thorax. However, within this work the body was divided into nine segments: thighs, knees, lower legs, arms and thorax (cf. Fig. 3). In addition, these segments have been approximated by frustums to match the non-cylindrical geometry of each segment individually by using the volume for calculating the body compositions of the segments. Each segment of the lower extremities consists of the tissues fat, bone and muscle. All segments have been simulated using FIT and an anatomical data set of a male human. The program used for this is CST EM Studio® from Computer Simulation Technology, Darmstadt, Germany. The data set is based on the Visible Human Data Set from the National Library of Medicine, Maryland, USA and provides resolutions from $[1 \times 1 \times 1]$ mm³ to $[8 \times 8 \times 8]$ mm³ (National Library of Medicine 2000). The origin of this data is Joseph Paul Jernigan, an executed prisoner, who has been frozen into gelatin after death. His frozen body was then cut into more than 1800 slices, so-called cryosections. These slices have been digitized by using MRI and CT with an image resolution of 4096×2700 pixels and a 24 bit color depth. At the moment, there are over 2000 licences sold in 48 countries regulating the usage of this data set. This work uses an anatomical data set created by MVR Studio GmbH, Loerrach, Germany. This data set already contains segmented voxels for several tissues. Since voxels contain no volume data for a certain tissue, the voxel data had to be converted into volume data. First, all tissues were converted into stereolitography (STL) files. STL data contains surface geometries composed of triangles, represented by polygon meshes, enclosing the different tissues. This enclosed area can be converted into volume data. That way, every tissue has been converted into a volume. By using boolean operations, all tissues have been combined to form the whole body. Therefore, one volume containing all tissues has been created and used as a fat meta structure into which all other tissues have been inserted by subtracting them from the meta tissue. Since the original voxel data contains no information about skin, it is not included in the volume data. To be able to include skin in the simulations, 5 mm skin tissue has been added at the electrode sites. After that, electrodes composed of an aluminium and a PEC (perfect electric conducting) layer have been added. The resulting volume model consumes 3.3 GB RAM.

The PECs have been connected to a fixed voltage. To assess the complex current, the current density over transversal faces has been integrated. Since the maximum wavelength is much higher than our measuring volume (cf. eq. 4), the low frequency electroquasistatic solver has been used.

$$c = \frac{\lambda}{f} = \frac{300000 \text{ km/s}}{5 \text{ MHz}} = 60 \text{ m} \tag{4}$$

This leads to a problem concerning the imaginary part of the complex current because the boundary conditions for electroquasistatic and magnetoquasistatic cases, respectively, are:

$$\frac{\partial \vec{D}}{\partial t} = 0 \quad \text{and} \quad \frac{\partial \vec{B}}{\partial t} = 0 \tag{5}$$

Figure 3.   Classical and used segmental subdivision.

The solution for this problem was to access the displacement current and assume a harmonic oscillation ($\vec{D} = |D| \cdot e^{j\omega t} \cdot \vec{e}_z$). Thus, the complex current could be calculated using the following formula:

$$\underline{I} = \int_A \left(\vec{J} + \frac{\partial \vec{D}}{\partial t} \right) d\vec{A} = \int_A \left(\vec{J} + j\omega \vec{D} \right) d\vec{A} \tag{6}$$

The material of the borders of the cubic simulation domain have been set to PEC. In addition, the interspace between simulation object and boundaries has been filled with vacuum (permittivity $\varepsilon_r = 1$, conductivity $\sigma = 0$). After having set the mesh density to create 2 million tetrahedrons, the simulation consumed 5 GB RAM and took 25 hours per sweep for the lower extremities. A whole body simultation would have taken 72 hours per frequency. Apart from knee, thigh and lower leg, knee-to-knee and foot-to-foot measurements have been simulated additionally as further scenarios. The impedances obtained by the simulations have been analyzed in Matlab® to compute extra- and intracellular resistances. The calculation frequency ranged from 5 kHz to 5 MHz, covering the measuring range of common BIS-analyzers.

For validation purposes, each segment has been measured at five male subjects with similar age (26–28 years) and Body Mass Index (21–25.5). The measurement device was a Xitron Hydra 4200 from Xitron Technologies, San Diego, USA. Standard adhesive electrodes from Fresenius Medical Care, Bad Homburg, Germany with a size of 19 mm × 80 mm have been used. All subjects were not allowed to drink and eat two hours before the measurements began. The subjects changed from an upright to a supine position remaining supine for 30 minutes so that a fluid shift from the lower extremities to the rest of the body occurred. To gather information about this shift, whole-body bioimpedance has been obtained after each segmental BIS measurement (Medrano 2009).

4 RESULTS

Before the human model has been built and used for simulation, the ability of the simulation program to simulate BIS has been tested by creating a capacitor with a muscle dielectric. This capacitor was intended to reproduce the measurements accomplished by Gabriel et al. who measured permittivity and conductivity of human tissues between 10 Hz and 1 GHz using network analysers. The results showed that the CST software could correctly reproduce the measurements by Gabriel et al. with, α, β and γ dispersion regions.

The model used for the simulation results is shown in Fig. 4.

The $\vec{D}$-Field results of a foot-to-foot measurement simulation are presented here by arrows floating through the body from the left foot to the right foot. Thickness and direction of the arrows symbolize the displacement current density and its direction.

Simulation results were analyzed using frequency response locus plots which resemble cole-cole curves (cf. Fig. 2, left), although the curve progression deviates slightly. Furthermore, the calculated values of the simulation results yield higher impedances in relation to

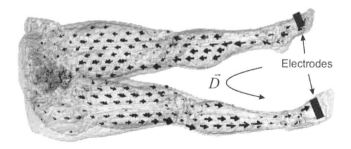

Figure 4.   Simulation model and results of a foot-to-foot measurement.

61

measured values found in literature and thus they have been corrected by a factor which has been calculated to divide the impedances (cf. Table 1).

To verify the simulations, segmental BIS measurements on five subjects ($n = 5$) have been carried out. The sequential whole-body BIS reference measurements, gathered to measure the fluid shift due to a position change, have been used for the correction of the intra- and extracellular resistances as shown in (Medrano 2009). An impedance shift towards higher impedances over 30 minutes was observed (cf. Fig. 5). It could be shown, that R_e rises while R_i decreases with time. The results of the segmental measurements revealed that the extra- and intracellular resistances of each of the six segments are related to the corresponding whole-body resistances and that the highest relative share is covered by whole extremities, the lowest by thighs. In addition, the sum of the resistances of segments forming the whole body (e.g. arm, leg and half of the thorax) equals approximately the resistivity of the whole body resistance.

Using the correction factors and a curve fit for the results of the simulations, measurements and simulations produced similar results (cf. Table 2).

Fig. 6 shows the corrected simulation results and its curve fit (dashed line) for a foot-to-foot simulation. Comparing this plot to the frequency locus plot of the measurement results (Fig. 7), one can see that both complex impedances lie within the same range.

Table 1. Comparison: R_E simulated and measured literature values.

Segment	$R_{e,sim}$ (Ω)	$R_{e,literature}$ (Ω)
Foot-to-foot	1915.9	243.2
Leg	1115.8	243.8
Knee-to-knee	252.9	54.3
Lower-leg	973.4	138
Thigh	238.4	27.1
Knee	238.4	78.7

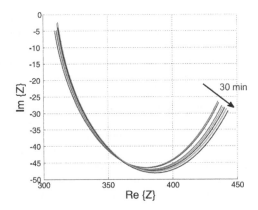

Figure 5. Hand-to-foot reference measurements.

Table 2. Comparison: R_E, R_I simulated and measured.

Segment	$R_{e,s}$ (Ω)	$R_{e,m}$ (Ω)	$R_{i,s}$ (Ω)	$R_{i,m}$ (Ω)
Foot-to-foot	479.85	418.79	642	855.28
Leg	276	229.8	333	411.07
Knee-to-knee	86.46	92.4	79.53	131.1
Lower-leg	176.97	135.01	222.38	212.1
Thigh	55.59	50.96	67.14	70.2
Knee	57.49	53.9	73.3	96.07

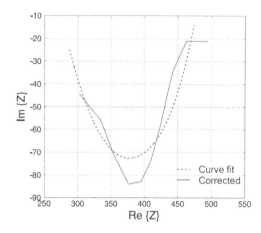

Figure 6. Simulation results, foot-to-foot.

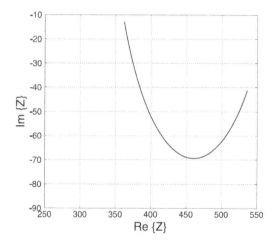

Figure 7. Measurement results, foot-to-foot.

5 DISCUSSION

The task of this work was to derive correlations of segmental BIS and whole-body BIS measurements and to compare them with simulations of segmental BIS measurements. Promising results have been obtained comparing measurements and simulations.

First, it hast been shown by segmental BIS measurements that intra- and extracellular resistances of segments correlate with whole-body resistances. The fluid shift from the lower extremities to the rest of the body during these measurements is explainable due to the position change of the subjects (Rush 2006). Second, results of the measurements reveal on the one hand that extremities have the highest relative share on the wholebody resistances which is explainable because they contain joints which are composed of low conducting bone tissue. On the other hand, the thighs contains much more good conducting material and thus they have the lowest influence on the whole-body resistance. Third, comparing measurements and simulations, this work also showed that simulations offer a reliable method for imitating processes in the body during BIS because they deliver the same dependencies of segments and the whole body.

However, some future work has to be done. Reasons for the simulation model to produce too high absolute impedances should be analyzed further to consider it in future simulations. The main reason could be the coarse model resolution of $8 \times 8 \times 8$ mm that leads to inaccuracies. During the conversion process small structures can be lost. Combined with some volumes such as small blood vessels (e.g. arterioles), which are too small to be considered in the model, these volumes are replaced by the meta volume fat tissue which is less conducting. Another known problem exists in this case because in the human body small blood vessels collapse after death due to pressure losses. This means that these volumes are not available, even in the finest voxel data. Another reason for higher impedances is the lack of skin in the original model. The added 5 mm skin volume does not reflect realistic impedances. Last but not least, a 2-point-measurement has been carried out in this work, which means that in fact, 8 cm tissue impedance should be subtracted for a 4-point-measurement from the simulated results.

Concerning the measurements, it is necessary to gather results from more subjects of all genders and ages to be able to generalize the findings of this work.

Considering that the intra- and extracellular resistances of segments are correlated to the whole-body resistance, it has been shown that it is possible to apply the body composition drawn from segments to the composition of the whole body. This leads to the possibility of using knee-to-knee measurements to monitor the body composition of geriatric patients. These measurements can be obtained e.g. from shorts worn by the patients containing integrated electrodes. The measured data could be exchanged via wireless networks to the attending doctor for a 24/7 supervision of the patient.

ACKNOWLEDGEMENTS

This work was funded by the German Federal Ministry of Education and Research within the program Micro Systems.

REFERENCES

Bracco, D. (1996). Segmental body composition assessed by bioelectrical impedance analysis and dexa in humans. *Journal of Applied Physiology 81*, 2580–2587.

Bundesministerium für Bildung und Forschung (2007). Gesundheitsforschung hautnah erleben. Website. http://www.bmbf.de/press/2171.php.

Carter, M. (2007). Estimation of Arm Muscle Mass in Hemodialysis Patients by Segmental Bioimpedance (SBIS) MRI and ^{40}K Techniques. In *IFBME Proceedings*.

Clemens, M. (2001). Discrete electromagnetism with the finite integration technique. *Progress In Electromagnetics Research 32*, 65–87.

Grimnes, S. and O.G. Martinsen (2000). *Bioimpedance and Bioelectricity Basics*. Academic Press.

Lavizzo-Mourey, R. (1987). Dehydration in the elderly: A short review. *Journal of the National Medical Association 79*, 1033–1038.

Lorenzo, A.D. (1997). Predicting body cell mass with bioimpedance by using theoretical methods. *Journal of Applied Physiology 82*, 1542–1558.

Matthie, J. (2005). Second generation mixture theorie equation for estimating intracellular water using bioimpedance spectroscopy. *Journal of Applied Physiology 99*, 780–781.

Medrano, G. (2009). Model-based correction of the influence of body position on continous segmental and hand-to-foot bioimpedance measurements. Submitted to: Medical & Biological Engineering & Computing.

National Library of Medicine (2000). The visible human project. Website. http://www.nlm.nih.gov/research/visible/visible_human

Nescolarde, L. (2007). Whole-body and thoracic bioimpedance measurement: Hypertension and hyper-hydration in hemodialyisis patients. In *Proceedings of the 29th Annual International Conference of the IEEE EMBS*.

Rush, E.C. (2006). Validity of hand-to-foot measurement of bioimpedance: Standing compared with lying position. *Obesity, a research journal 14*, 252–257.

Salmi, J. (2003). Body composition assessment with segmental multifrequency bioimpedance method. *Journal of Sports Science & Medicine 2*, 1–29.

Statistisches Bundesamt (2007). Entwicklung der Lebenserwartung Fernere Lebenserwartung im Alter von 60 Jahren, 1901 bis 2050. Website. http://www.bpb.de/files/XH3MK2.pdf.

Thomas, B.J. (2003). A comparison of the whole-body and segmental methodologies of bioimpedance analysis. *Acta Diabetologica 40*, 236–237.

Wabel, P. (2009). Importance of whole-body bioimpedance spectroscopy for the management of fluid balance. *Blood Purification, 27*, 75–80.

Zhu, F. (2006). Segment-specific resistivity improves body fluid volume estimates from bioimpedance spectroscopy in hemodialysis patients. *Journal of Applied Physiology 100*, 717–724.

Lecture Notes on Impedance Spectroscopy – Kanoun (ed)
© 2011 Taylor & Francis Group, London, ISBN 978-0-415-68405-7

Design and evaluation of a portable device for the measurement of bio-impedance-cardiograph

Qinghai Shi
Chair for Measurement and Sensor Technology, Chemnitz University of Technology, Chemnitz, Germany

Andreas Heinig
Lifetronics, Fraunhofer Institute for Photonic Microsystems, Dresden, Germany

Olfa Kanoun
Chair for Measurement and Sensor Technology, Chemnitz University of Technology, Chemnitz, Germany

ABSTRACT: Electrical impedance of biological matter is known as electrical bio-impedance or simply as bio-impedance. Bio-impedance devices are of great value for monitoring the pathological and physiological status of biological tissues in clinical and home environments. The technological progress in instrumentation has significantly contributed to the progress that has been observed during the last past decades in impedance spectroscopy and electrical impedance cardiograph. Although bio-impedance is not a physiological parameter, the method enables tissue characterization and functional monitoring and can contribute to the monitoring of the health status of a person. In this paper an inexpensive portable multi frequency impedance cardiograph device based on impedance spectroscopy technique has been developed. By means of this system the basic thoracic impedance range and the heart-action-caused changes of impedance can be measured and the hemodynamic parameters of the heart function can be determined. This system has small size and low current consumption. The impedance cardiograph signals of the electrodes configuration by Sramek, Penney and Qu in this work was measured; compared and summarized. The differences of the measuring method, the schematic circuit diagram, the measurement results and area of application between impedance cardiograph and impedance spectroscopy were discussed and compared. The performance of this sensor-system was evaluated.

Keywords: impedance cardiograph (ICG); impedance spectroscopy (IS); electrocardiography (ECG); electrical bio-impedance of the chest; cardiac outputs; hemodynamic; non-invasive diagnostic

1 INTRODUCTION

Evaluation of the hemodynamic parameters of a patient has always been a subject of interest to clinicians. It has been difficult to capture hemodynamic parameters and invasive methods have been used to measure them. These invasive techniques are expensive, time-consuming, demand complicated equipment and trained staff and are not always possible to use because of the condition of the patient, which maybe either too serve or else too good to run the risks associated with invasive techniques (Sodolski and Kutarski 2007).Therefore they are not suitable for long time or repeated measurements because of their invasive nature. Several non-invasive techniques capable of monitoring cardiac activity were in past two decades developed. They are ultrasound Doppler, magnetic resonance imaging, and impedance cardiograph (ICG). However, all these techniques except ICG are not suitable for long-term continuous monitoring of cardiac activity. Impedance cardiograph is a simple, inexpensive, and non-invasive method for hemodynamic parameters measurements. The method is based on changes in the electrical resistance of the chest during heartbeat (Strobeck and Silver 2004). Previous Studies suggest that various physiological variables contribute to the impedance changes. The most commonly mentioned contributors are:

Enlargement of the volume of aorta, enlargement of the volume of the blood in the pulmonary circulation and laminar blood flow in the large vessels (Osypka and Bernstein 1999). It is generally accepted that ICG originates from a combination of blood volume change, rearrangement of current conductors, redistribution of current density, and resistivity change of flowing blood within the measuring object. ICG was developed into clinical practice by Kubicek and Colleagues using their proposed equation and band electrode array [Fig. 1]. However, the band electrodes are not practical for use. The two electrodes around the neck can produce an annoying, or even a choking sensation. This maybe cause increased apprehension in some patients; which could be harmful in patients with cardiac disease. Furthermore, it is difficult to place the full band electrodes on patients with chest burns or on patients recovering from cardiac surgery who have incisions and dressings or tubes and lines where the electrodes need to be applied (Kubicek 1970). New electrode arrays were introduced using disposable spot electrodes. In modern ICG systems electrode configurations by Sramek, Penney and Qu are used (Sramek 1981) (Penney and Wheeler 1985) (Penney 1986). In electrode configuration by Sramek four electrodes are used to deliver the electrical current, which are known as "current electrodes". Another four electrodes are used to measure the voltage changes, which are known as "voltage electrodes". The electrodes are positioned symmetrically on both sides of the patient's neck and the chest [Fig. 1B]. An alternating current flows along the current electrodes through the measured body segment and the voltage electrodes measure the voltage changes. For the electrode configuration by Penney the electrode array, as illustrated in [Fig. 1C], uses for ECG spot electrodes. Two are placed at the base of the neck, separated by 6 cm and centered about the prominence of C-7. Two more electrodes are placed below the heart on the left anterolateral chest surface. One is placed at the end of the ninth intercostals space, near the mid-clavicular line. The other is 8 cm far away from the first, in the tenth intercostal space, near the mid-axillary line. The current is passed between the electrode on the right of the neck and that at the end of the ninth intercostals space. The voltage difference is measured between the remaining electrode pair (Penney and Wheeler 1985). Qu et al. developed a new plot electrode configuration. One voltage electrode is placed 4 cm above the clavicle on the anterior surface of the neck; a second voltage electrode placed over the sternum at the lever of the fourth rib; a current electrode placed at the lever of the fourth vertebrae on the posterior surface of the neck; a second current electrode placed at the lever of the ninth thoracic vertebrae on the back. This plot electrode configuration yielded improved signal quality because the placement of sensors along the midline minimized movement and breath artifacts during exercise (Qu 1986) [10]. The impedance cardiograph signals

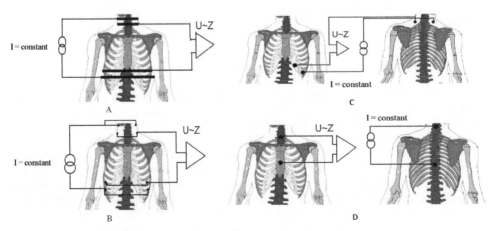

Figure 1.   A: band electrode configurations by Kubicek; B: 8 spot electrode configurations by Sramek; C: spot electrode configurations by Penney; D: spot electrode configuration by Qu.

of the electrodes configuration by Sramek, Penney and Qu in this work was measured; compared and summarized.

2 METHODS

The hardware module in this work is based on the tradition method where a constant current $I(t)$ (approximately 1 mA at 100 kHz) fed into current electrodes [Fig. 2] and the resulting voltage $U(t)$ is measured through inner voltage electrodes, thereby allowing the calculation of the electrical bio-impedance $Z(t)$:

$$Z(t) = \frac{U(t)}{I(t)} \tag{1}$$

To realize this simple method, the following issues have to be kept in mind: the current fed into the patient needs to be a well defined constant signal with fixed magnitude, time and frequency. A direct digital synthesis (DDS) IC is used to generate the sine wave signal. This sin wave voltage can be fed into the patient through the ECG Electrodes. The resulting patient voltage is detected by an instrumentation amplifier and the current is measured by the other instrumentation amplifier at the same time. Then the gain phase detector can be used to detect the magnitude and phase of the bio-impedance of the selected segments. The magnitude of bio-impedance consists of two important signals: the base thoracic impedance and thoracic impedance changes. There are some noises in this signal: the impedance changes because of breath and motion which should be eliminated. Then a differentiator is used to derivate the thoracic impedance changes. All of the analog signals are sent to analog to digital convert of the microprocessor MSP430f1611 and over there digital signal processing for calculation of hemodynamic parameters can be programmed. MSP430f1611 is the microcontroller configurations with two build-in 16-bit timers, the fast 12-bit analog to digital converters, dual 12-bit digital to analog converters, universal serial synchronous/asynchronous communication interfaces (USART), I2C, DMA and 48 I/O pins.

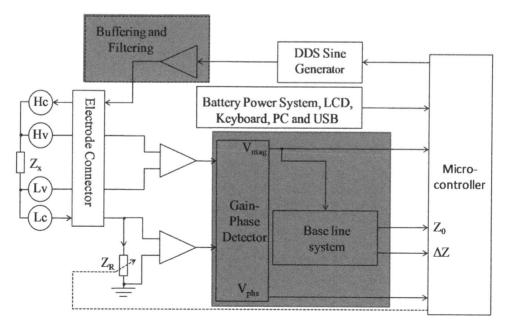

Figure 2. The block diagram of ICG System.

This system can be considered as amplitude modulation in communication system for the information transmission [Fig. 9]. The stimulated signal is considered as carrier wave. The measured impedance signals can be considered as the original signals. Therefore the output signals of this system can be considered as the modulated signal. The output signals can be separated into the different parts with the gain phase detector and base line system. The measured impedance consists of three different parts: the base thoracic impedance Z_0, heart action caused impedance changes ΔZ_{heart} and breath caused impedance changes ΔZ_{heart}, the base line system can be used to separate these three different parts from the measured signal. Like we know that the heart beat frequency f_1 is from 1 to 2.5 Hz and the breath frequency f_2 is from 0.25 to 0.43 Hz. In 1 the H is the amplitude of heart action caused of impedance changes ΔZ_{heart} and B is breath caused impedance changes ΔZ_{breath}. The noise comes from the system and the stimulus signal.

$$Z = Z_0 + \Delta Z_{heart} + \Delta Z_{breath} \tag{2}$$

$$Z = Z_0 + H \cdot \sin(2 \cdot \pi \cdot f_1 \cdot t + \varphi_1) + B \cdot \sin(2 \cdot \pi \cdot f_2 \cdot t + \varphi_2) + noise \tag{3}$$

An automatically balancing circuit is used to detect the thoracic impedance which consists of two low pass filters with the cut off frequency 22 Hz and 0.7 Hz and a differential amplifier [Fig. 10]. With the low pass filter with cut off frequency 22 Hz the noise signals can be eliminated. With another low pass filter with cut off frequency 0.7 Hz the heart action caused impedance changes is ejected and we can get the base thoracic impedance and breath caused impedance changes. Then with the digital signal processing these two signals can be separated and saved. A differential amplifier is used to acquire the heart action caused impedance changes. Therefore these three different parts can be separated.

3   RESULTS AND ANALYSIS

In this capital the electrode configurations by Sramek, Penney and Qu are measured and compared with this ICG System. As illustrated in [Fig. 3] the base thoracic impedance Z_0 is 28Ω (2.8V, $G = 100$). The thoracic impedance changes is minimal 0.5Ω and maximal 0.8Ω which is ca. 3% of the base thoracic impedance (0.6V to 0.96V, $G = 1200$). The four current spot electrodes are symmetrically placed on both side of the neck and chest. Therefore that generates homogeneous electrical field and the impedance of electrode can be neglected so

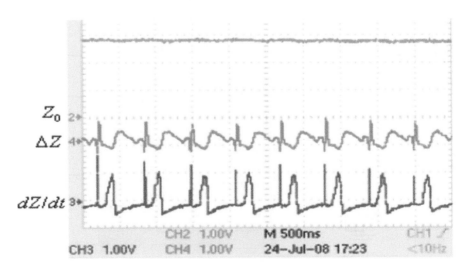

Figure 3.   Results of Sramek electrode configuration.

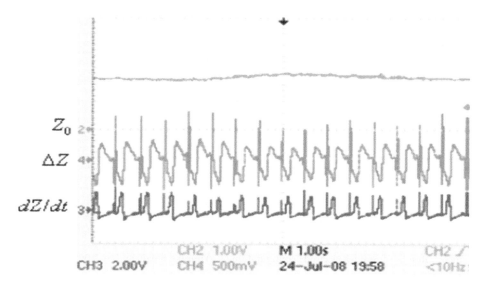

Figure 4.   Results of Penney electrode configuration.

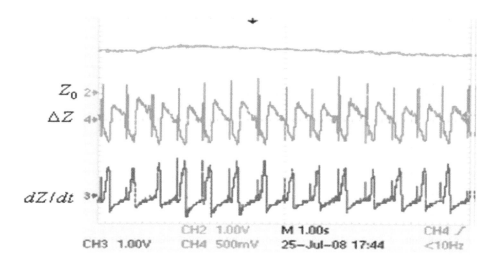

Figure 5.   Results of Qu electrode configuration.

that these can improve the accuracy of measurement. The reproducibility of measurement of this electrode configuration is best than the other plot electrode configurations. The base thoracic impedance Z_0 is 18Ω ($1.8V$, $G = 100$) [Fig. 4]. The thoracic impedance changes is minimal 0.5Ω and maximal 0.8Ω (0.6 to $0.96V$, $G = 12000$) and is ca. 4% of the base thoracic impedance. The base thoracic impedance by this configuration is smaller than by Sramek. As showed in Fig. 5 the base thoracic impedance Z_0 is 18Ω as same as the electrode configuration by Penney ($1.8V$, $G = 100$). The thoracic impedance changes is minimal 0.6Ω and maximal 0.9Ω (0.72 to $1.1V$, $G = 1200$) and ca. 5% of the base thoracic impedance which is greater than another electrode configuration. In this work the influence of breath for the thoracic changes by these three spot electrode configurations is evaluated. As illustrated in Fig. 6–8 we can make a conclusion, that Qu spot electrode configuration has the best signal noise ratio than another two configurations. Because from Fig. 8 we can see that the thoracic impedance changes has the constant basis line.

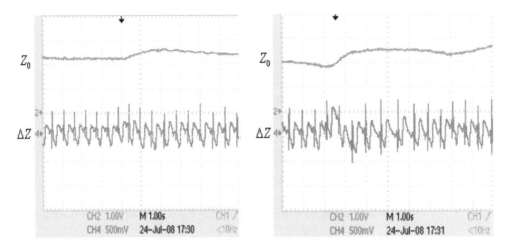

Figure 6. Results of Sramek electrode configuration by a) normal breath and b) depth breath.

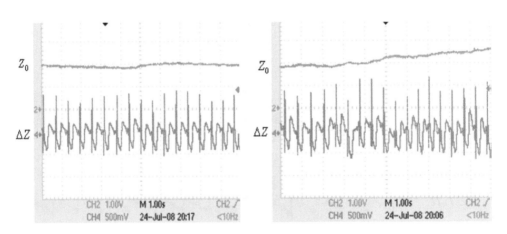

Figure 7. Results of Penney electrode configuration by a) normal breath and b) depth breath.

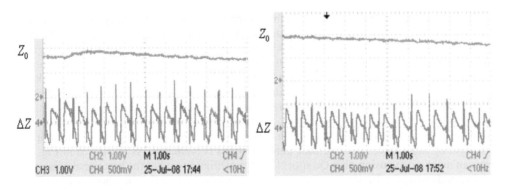

Figure 8. Results of Penney electrode configuration by a) normal breath and b) depth breath.

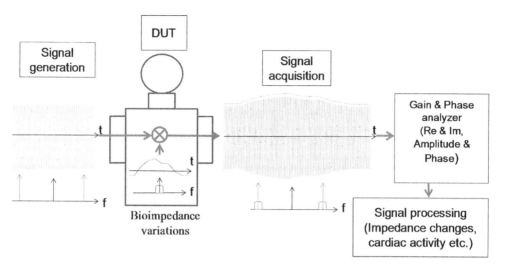

Figure 9.   Structure diagram of a Bio-impedance measurement system.

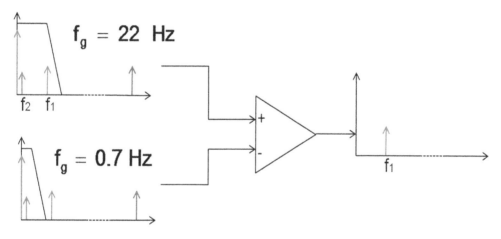

Figure 10.   Differential amplifier.

4   CONCLUSIONS

The electrodes configurations and signal processing play an important role in signal reproducibility. In this study the results by different electrodes configuration (Sramek, Penney and Qu) are measured and compared with the same circuit. The results of electrodes configuration by Sramek are similar as band electrodes configuration. But this electrodes configuration is sensitive to breath and motion. The results of electrodes configuration by Qu have best accuracy and the breath and motion have no effect on the results of this electrodes configuration. In this work a basis line system is used to eliminate the effect of breath and motion on the results. With this system the section of heart-action caused changes of impedance and of breath caused changes of impedance can be separated. That means, the noise of breath caused changes of impedance not only can be eliminated from the basic thorax impedance and heart-action caused changes of impedance but this signal can be also measured in this system. In the future, more advanced techniques such as adaptive filter, specific model and digital signal processing for calculation of hemodynamic parameter can be implemented for detection to increase the accuracy and reliability. The SD card can be used in the future to storage the large data. A fewer components will be hoped to use and therefore as few as possible power can be used that a long time monitoring system can be achieved in the future.

REFERENCES

Kubicek, W.G. and Patterson, R.D. (1970). Impedance cardiography as a non-invasive method of monitoring cardiac function and other parameters of the cardiovascular system. *Ann. N. Y. Acad. Sci.*, 724–732.

Osypka, M. and D. Bernstein (1999). Electrophysiological principles and theory of stroke volume determination by thoracic electrical bio-impedance. *AACN Clin Iss. Vol. 10*, 385–395.

Penney, B. (1986). Theory and cardiac applications of electric impedance measurements. *Critical Reviews in Biomedical Engineering Vol. 3*, 227–281.

Penney, B.C., N.A.P. and H.B. Wheeler (1985). Simplified electrode array for impedance cardiography. *Medical and Biological Engineering and Computing Vol. 23*, 1–7.

Qu, M. and Zhang, Y.W. J.T.W. (1986). Motion artifact from spot and band electrodes during impedance cardiography. *IEEE Transactions on Biomedical Engineering BME 33*, 1029–1036.

Sodolski, T. and A. Kutarski (2007). Impedance cardiography: A valuable method of evaluating hemodynamic parameters. *Journal Vol. 14. No. 2*, 115–126.

Sramek, B.B. (1981). Non-invasive technique for measurement of cardiac output by means of electrical impedance. *Tokyo. Proc. of the 5th ICEBI*, 39–42.

Strobeck and M. Silver (2004). Beyond the four quadrants: the critical and emerging role of impedance cardiography in heart failure. *Congest Heart Fail Vol. 10*, 1–6.

Lecture Notes on Impedance Spectroscopy – Kanoun (ed)
© *2011 Taylor & Francis Group, London, ISBN 978-0-415-68405-7*

Implementing bio impedance spectroscopy for application specific multichannel measurement & diagnostic solutions

S. Wegner
Sciospec Scientific Instruments, Schmölen, Germany

ABSTRACT: Todays conventional setups for the instrumentation of bioimpedance spectroscopy (BIS) are in most cases not fit for application specific implementation. Despite its growing importance in biotechnology, medicine and virtually all related fields of research as well as the still largely unexploited potential in diagnostic use, hardly any hardware specially designed for electrical bioimpedance spectroscopy is available today. An innovative approach based on a modular scalable electronic concept has been developed enabling cost effective, fully adaptable single and multichannel, precision and/or high throughput applications for cell impedance spectroscopy.

The following text gives a brief review of the concepts involved and shows how this approach can help realize applications as suggested by current trends.

Keywords: EIS, BIS, electrical bio impedance, cell-impedance, spectroscopy

1   INTRODUCTION

Conventional instrumentation setups for electrical bio impedance spectroscopy generally either aim towards precision laboratory operation or high throughput, low precision indicator extraction. The concepts behind the technical realization could not be more different. While in the lab application a wide application span has to be covered, requiring high resolution, bandwidth and range, the high throughput application needs to be efficient both in cost and performance, resulting in a highly specialized, but low resolution measurement of a few frequencies in the spectrum and coarse parameter fitting or indicator detection. Both approaches share one major disadvantage: They are generally not easily adaptable for new application specific implementations. The new modular approach was developed to tackle exactly this problem.

2   CONCEPT OVERVIEW

In order to adapt one concept to basically any possible application of BIS it is necessary to enable both high resolution, range and precision, as well as cost and time effective multichannel realizations. For all steps in the instrumentation chain easily interchangeable modules have been developed, e.g. either analog lock-in or direct sampling techniques are used depending on the application's SNR, frontends can be easily trimmed for the exact bandwidth required, adaptive gain settings and signal compression enable high dynamic ranges and for cost sensitive high channel count applications a fully integrated frontend solution is available. The heart of the electronic system is a FPGA based controlling platform. Depending on the specific demands this main unit either functions plainly as communication junction between the system and the PC based controlling software or incorporates process control, data processing and analysis, error correction as well as data management and storage, making the system self contained. Combinations of the frontend choices result in an application specific channel design that now can be completed with a memory management node enabling basically any desired number of channels in one design without reengineering any part of the signal chain.

For the realization of a new application of BIS the best suited modules can now be chosen depending on the specific parameters and demands and with minimal effort and no reengineering a new hardware can be assembled. This does not only minimize the technical difficulties with new implementations, but also the costs for prototyping.

3 POSSIBLE APPLICATIONS

A fair number of studies based on electrical impedance spectroscopy have been published ranging from characterization of cell types, over bioreactor process control to cell parameter extraction and studies on the reaction of cell properties to changes in environment (e.g. pharmacological screening). With suitable means of implementing BIS in ready to use cost effective applications it is now possible to benefit from all this research and ultimately translate the efforts already made into process control, monitoring or diagnostic solutions.

Examples of applications include electrochemical characterization, cell toxicity, blood characterization (e.g. hematocrit, blood oxygen and glucose levels or cell differentiation), patch clamp based single cell analysis or general process monitoring in bioreactors (cell growth and adherence) as well as environmental process and parameter monitoring. For most of these applications laboratory based setups and corresponding results have already been described in according publications. As for the interface between cell (single cell, cell-suspension, mixed volumes, tissue etc.) and the measurement electronics many possibilities are available, including multi electrode arrays, multi well plates, patch clamp (planar or pipette) configurations, needle electrodes, micro fluidic channels with electrodes etc. The diversity of electrode configurations ensures a suitable fit for any BIS application that comes to mind.

4 CHANCES AND TRENDS

The trend towards system biology—the systematic approach to understanding and modelling biologic systems—demands for efficient ways to gain insight into these systems. Using methods with sparse information output might facilitate the mathematical effort for modelling the system, but in order to create more accurate models and better understand the system, more data has to be collected in a way, that can be used for the modelling task at hand. The combined measurement of high resolution, precise impedance spectra with other electrophysiological or optical techniques opens up entirely new possibilities of modelling biological cells.

In addition to the above mentioned combination of different electrochemical analyses, the application specific implementation of BIS offers the possibility to extend the impedance spectrum analysis in several ways. Fast acquisition times enable good time resolution for process control (spectrogram analysis) and the variation of input signals allows for nonlinear spectroscopy and thus nonlinear modelling.

One major disadvantage of many analytic applications is the fact that the cells have to be disposed after the test and cannot be used for further analysis. This also means that in-process measurements are very limited. The main reason for this is that up to now almost exclusively high throughput applications have been realized, making multi well or multi electrode array assemblies necessary. With cost efficient ways to create application specific solutions it is now possible to create small volume applications, targeting flow-through, in-system setups for non-invasive, online measurements.

5 CONCLUSION

Bio impedance spectroscopy has proven to be a powerful analytical tool in the laboratory for decades and even though many studies based on BIS have been done, it still shows largely unexploited potential and application specific implementations are rare up to today. Being label free, fully automatable and reliable BIS is suitable for many applications ranging from laboratory research, over small diagnostic solutions to industrial high throughput applications.

The main advantages of the new concept include scalability achieved through the modular approach, fast prototyping, diversity of cell interface possibilities, capability of online parameter extraction and flow-through setups, as well as the possibility to create ready to use BIS applications that are easy to operate and have small dimensions, enabling both bench top and portable solutions.

With the new modular approach towards the implementation of bio impedance spectroscopy systems it is now possible to easily prototype new applications and ultimately carry bio impedance spectroscopy out of the laboratory into practical usage.

REFERENCES

Ayliffe, H.E., A.B. Frazier, and R.D. Rabbitt (1999). Electric impedance spectroscopy using microchannels with integrated metal electrodes. *IEEE Journal of Microelectromechanical Systems 8*, 50–56.

Bao, J.-Z., C.C. Davis, and R. Schmuklert (1992). Frequency domain impedance measurements of erythrocytes constant phase angle and impedance characteristics and a phase transition. *Biophysical Journal 61*, 1427–1434.

Coverdale, R., H.M. Jennings, and E.J. Garboczi (1995). An improved model for simulating impedance spectroscopy. *Computational Materials Science 3*, 465–474.

Gawad, S., M. Wthrich, L. Schild, O. Dubochet, and P. Renaud (2001). On-chip impedance-spectroscopy for flow-cytometry using a differential electrode sensor. In *Transducers 01: Eurosensors XV, Digest of Technical Papers Vol. 1 and 2*.

Orazem, M.E. and B. Tribollet (2008). *Electrochemical Impedance Spectroscopy*. Number ISBN 978-0-470-04140-6. Wiley.

Pfützner, A., A. Caduff, M. Larbig, T. Schrepfer, and T. Forst (2004). Impact of posture and fixation technique on impedance spectroscopy used for continuous and noninvasive glucose monitoring. *Diabetes Technology & Therapeutics 6*, 435–441.

Schanne, O.F. and E. Ruiz P.-Ceretti (1978). *Impedance Measurements in Biological Cells*. Wiley Interscience.

Measurement methods

Lecture Notes on Impedance Spectroscopy – Kanoun (ed)
© *2011 Taylor & Francis Group, London, ISBN 978-0-415-68405-7*

Chirp pulse excitation in the impedance spectroscopy of dynamic subjects—Signal modelling in time and frequency domain

M. Min & R. Land
Department of Electronics, Tallinn University of Technology, Tallinn, Estonia

P. Annus & J. Ojarand
Department of Sensing and Communication, Competence Centre ELIKO, Tallinn, Estonia

ABSTRACT: Wideband excitation enables to perform the fast spectroscopy, that is, to cover a wide frequency range during a short time interval. It extends the measurement system to provide spectral information matched with the dynamics of the subject under study. Power spectrum of the chirp pulse excitation covers nearly the same frequency range in despite of different duration of pulses in time. The paper describes the results of the study of spectral properties of chirp signals depending on their duration.

Keywords: wideband excitation, impedance spectroscopy, chirp pulse, power spectrum, dynamic processes

1 INTRODUCTION

The simplest wideband excitation is a short rectangular pulse (Pliquet, Gersing, and Pliquett 2000), either unipolar or bipolar one (10 and 20 μs pulses in Figs. 1a and b). The problem is that only about 65% of generated energy falls into the useful bandwidth BW = 44 kHz (at −3dB or 0.707 RMS level corresponding to −6dB or 0.5 power level), whereby the RMS spectral density reduces to zero at 100 kHz (Figs. 2a and b). The spectrum reduces 20 $^{dB}/_{decade}$ outside the BW, also the crest factor (ratio of the peak and RMS values) of the signal is high and affects the matter under test. The 10 μs chirp pulse in Fig. 3 a covers the excitation bandwidth BW = 100 kHz and its RMS spectral density is almost flat within the BW (Fig. 3 b). About 80% of generated energy lies in the useful BW.

The chirp pulse in Fig. 3a is comparable to the rectangular pulse in Fig. 1a. This chirp pulse (Fig. 3a) can be described as

$$ch(t) = \sin\left[2\pi\,(2B/T)\cdot t_2/t\right] \tag{1}$$

where $0 \prec t \prec T_{exc}$ and duration $T_{exc} = T/2$ of the chirp pulse (1) is equal to half-cycle of sine wave. The chirp rate $\frac{B}{T_{exc}}$, Hz/s, corresponds to the excitation bandwidth BW = 100 kHz (Fig. 3 b) covered by the chirp pulse spectrum during one half-cycle $T/2 = 10$ μs of sine function (1). A bipolar chirp pulse with duration of one full wave cycle $T = 20$ μs in Fig. 4a corresponds to the rectangular excitation in Fig. 1b. Its RMS spectrum is shown in Fig. 4b.

2 CHIRP ENERGY AND IMPEDANCE DYNAMICS

An outstanding property of chirp function is that the useful excitation bandwidth BW can be set independently on duration of the chirp pulse when choosing appropriate chirp rate B/T_{exc} (1). In Fig. 5 a chirp pulse is shown consisting of 10 full cycles with duration

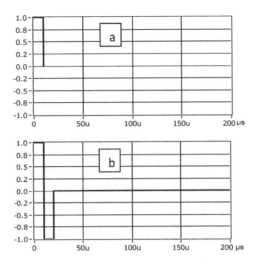

Figure 1. Unipolar 10 μs (a), and bipolar 20μs (b) rectangular pulses.

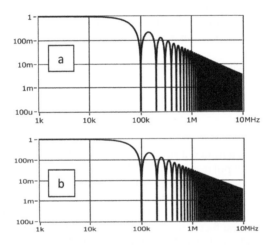

Figure 2. Density of RMS spectra of the unipolar (a) and the bipolar (b) rectangular pulses shown in Fig. 1a and b.

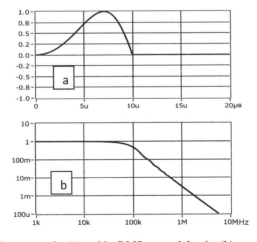

Figure 3. Half-cycle sine wave pulse (a) and its RMS spectral density (b).

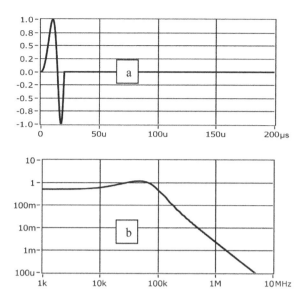

Figure 4. Full-cycle sine wave chirp (a) and its RMS spectral density (b).

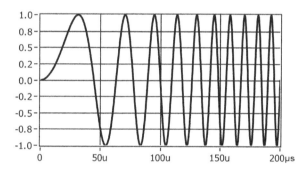

Figure 5. A chirp pulse consisting of 10 cycles, $T_{exc} = 200$ μs.

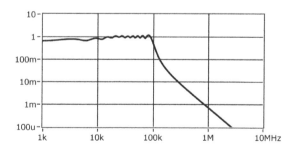

Figure 6. RMS spectral density of the chirp pulse consisting of 10 cycles, $T_{exc} = 200\mu$s.

$T_{exc} = 10T = 200\mu$s, the RMS spectrum density of this signal is depicted in Fig. 6. In Fig. 7 is given the RMS spectrum of the chirp with duration $T_{exc} = 10^5 T = 2$s.

Excitation energy depends proportionally on duration of the excitation pulse T_{exc}. Therefore, it is reasonable to use longer excitation pulses for obtaining a better signal-to-noise ratio. Besides other requirements, the main limiting factor is a speed of impedance variations (dynamics). Thanks to specific properties of the chirp function, it possible to match the needs for bandwidth, time, signal-to-noise ratio and dynamic requirements when implementing the

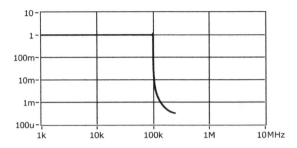

Figure 7.   RMS spectral density of the chirp pulse consisting of 10^5 cycles, $T_{exc} = 2s$.

impedance spectroscopy in lab-on-chip type analyzers (Min, Pliquett, Nacke, Barthel, and Annus 2008; Sun, Gawad, Bernabini, Green, and H. 2007) and in medical devices (Nahvi and Hoyle 2008).

ACKNOWLEDGEMENTS

The authors thank their colleagues from the EU FP6 Marie Curie ToK project no. 29857 InFluEMP workgroup at the Institut fr Bioprozess—und Analysenmesstechnik in Heilbad Heiligenstadt, Germany, for the design of microfluidic system and making it available for experiments. Special thanks belong to Dr. Uwe Pliquett for his guiding discussions on the principles of impedance spectroscopy.

REFERENCES

Min, M., U. Pliquett, T. Nacke, A. Barthel and P. Annus (2008). Broadband excitation for short-time impedance spectroscopy. *Physiological Measurement 26*, 185–192.

Nahvi, M. and B.S. Hoyle (2008). Wideband electrical impedancetomography. *Measurement Science and Technology 19*, 1–9.

Pliquet, U., F. Gersing and F. Pliquett (2000). Evaluation of Fast Time-domain Based Impedance Measurements on Biological Tissue—Beurteilung schneller Impedanzmessungen im Zeitbereich an biologischen Geweben. *Biomedizinische Technik/Biomedical Engineering 45*, 6–13.

Sun, T., S. Gawad, C. Bernabini, N.G. Green and M.H. (2007). Broadband single cell impedance spectroscopy using maximum length sequences: theoretical analysis and practical considerations. *Measurement Science and Technology 18*, 2859–2868.

Lecture Notes on Impedance Spectroscopy – Kanoun (ed)
© 2011 Taylor & Francis Group, London, ISBN 978-0-415-68405-7

Measurement concept for broadband impedance spectroscopy analysers for process applications

Y. Zaikou, A. Barthel, T. Nacke & U. Pliquett
Institute for Bioprocessing and Analytical Measurements e.V., Heilbad-Heiligenstadt, Germany

M. Helbig & J. Sachs
Electronic Measurement Research Lab, Ilmenau Technical University, Ilmenau, Germany

J. Friedrich & P. Peyerl
MEODAT GmbH, Ilmenau, Germany

ABSTRACT: In this paper, a modular concept for measuring broadband impedance spectroscopy is introduced with a detailed description and measurement examples.

Keywords: modular concept, crest factor, M-sequence

1 INTRODUCTION

Most commercial measurement devices in impedance spectroscopy are impedance analyzers that utilize the 'frequency sweep' concept for sensing the properties of interest across a wide frequency range. These are 'high-end' measurement systems, highly versatile and precise; they employ measurement principles that guarantees enormous data quality. However, this very principle does not allow the development of a small lightweight system (which is a prerequisite for many non-laboratory applications). Moreover, the requirement of highly trained personal (which arises due to the high flexibility of analyzers) and the ultimate parameter—cost of the system—often limits the acceptance of such devices.

In practical applications, there is a demand for robust, reliable and affordable instrumentation. Moreover, additional requirement for data processing arises: it should be automated and simple to use, so that the application of impedance measurement does not require highly trained technicians.

2 MEASUREMENT CONCEPT

The system should be capable of operating across a wide frequency range and flexible with respect to the interface and the software to cover as much practical applications as possible. This leads to the idea of the modular concept with basic units and special modules (see Fig. 1).

While the basic units are necessary for all applications, the special modules are interchangeable and they can be adapted to the material under test (MUT). The general purpose unit (GPU) controls data acquisition; it is used for any advanced data post-processing required for a particular application. This unit can be used to control a number of additional sensors and capture data in several user programmable channels supporting the high versatility of the entire system. In typical applications, it is convenient to use a Laptop-PC as GPU. The data processing unit (DPU) is the heart of the whole measurement system. The DPU is able to generate stimulus signal for low-frequency AU and to capture the data in two parallel measurement channels. Moreover, it can be used for critical data analysis (for some application this makes GPU redundant which simplifies the system and makes it cheaper).

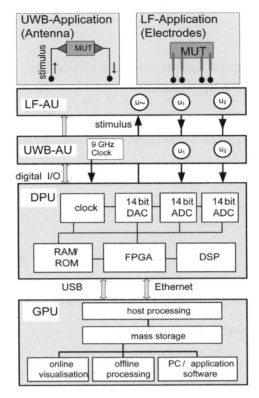

Figure 1.   Structure of measurement system for broadband impedance spectroscopy.

The application unit (AU) is available as: a) a low frequency application unit (LF-AU) up to 10 MHz and b) an ultra wideband application unit (UWB-AU) up to 5 GHz (see Fig. 2 and Fig. 3). In LF-AU electrical current and voltage drop are measured by means of electrodes in order to estimate complex impedance. Differential signals are used to minimize the influence of transmission lines. The measurement concept allows a stimulus signal of arbitrary shape for this unit and thus, a multisine signal with minimum crest factor is usually chosen. In short, the choice of such stimulus allows us to take advantage from the same source as laboratory impedance analyzers, since similar approach is used in those devices.

The low crest factor of the sine waves promotes the handling of signals rich in energy resulting in high Signal to Noise and Distortion (SINAD)-values. Moreover, signal sources provide for stable operational conditions so that effective methods can be applied to remove systematic errors.

In UWB-AU, however, completely different approach is used in order to meet the system requirements. A digital shift-register of order m with appropriate feedbacks provides Maximal Length Sequence (an M-Sequence) having a bandwidth up to several GHz (depending on clock-rate f_c). M-sequence is a periodical signal with all possible combinations of binary numbers of order m (excluding all zeros) inside one period. This signal stimulates the MUT (by means of UWB antennas, coaxial probe, etc). In contrast to the impulse technique (and similar to frequency sweep approach), M-Sequence distributes its energy over the complete measurement period. That is, for maximal appropriate value of signal amplitude determined by electronic components resulting power and Signal to Noise Ratio (SNR) is higher than that in the impulse technique and it is comparable to those of frequency-sweep devices (see (Sachs, Peyerl, and Zetik 2003) for thorough comparison of M-sequence devices with other methods). Signals with low amplitudes are easy to handle from hardware point of view and they allow a monolithic circuit integration resulting in an improved RF behavior (realization of critical system components on a circuit is shown in Fig. 3). The envelope of the power

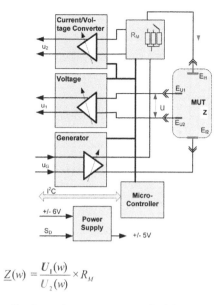

$$\underline{Z}(w) = \frac{U_1(w)}{U_2(w)} \times R_M$$

Figure 2. Low-frequency application unit, measurement principle.

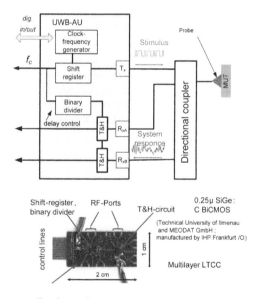

Figure 3. High-frequency application unit, measurement principle.

spectrum of the M-Sequence has sinc-squared shape with zeros at a multiple of the clock rate fc. It can be seen in Fig. 4 that roughly 80% of the signal energy is concentrated to the frequency band below $f_c/2$. Therefore, cutting the frequency band of interest at $f_c/2$ does not lead to dramatic loss in performance. Respecting the Nyquist theorem leads again to an (equivalent) sampling rate f_s which is equal to the clock rate $f_s = f_c$ i.e. one sample per chip of the M-Sequence. The period of an M-Sequence always includes $N = 2n - 1$ chips in which n is the length of the shift register. Since there is no number N of such kind that could be divided by 2, a binary divider of arbitrary length can be used for generating sampling events.

The steep flanks of the divider as well as the accuracy of the temporal sample spacing contribute positively to the stability of entire measurement system. A sampling event is generated only after the binary divider has run completely through all its internal states. Moreover,

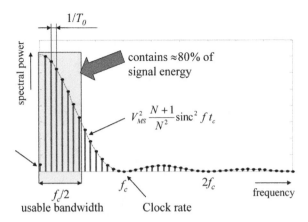

Figure 4. Power spectrum of idealized M-sequence.

this sampling approach creates the opportunity to adapt the speed of data gathering to the actual needs of the particular measurement task. In highly dynamic applications, short binary dividers are used so that more than one data sample per period is captured. In the end, one limiting factor concerning the measurement speed is the digital data processing which has to follow the data stream. In more stationary measurement scenarios, longer binary dividers are used resulting in reduced requirements concerning the processing speed.

3   MEASUREMENTS AND RESULTS

Three examples of measurements are given below to demonstrate the viability of measurement concept. More measurement results both for LF AU and UWB AU (calculation of dielectric constant of liquids in wide frequency range, determination of water content in living tissue, etc.) are described in (Nacke, Barthel, Friedrich, Helbig, Sachs, Peyerl, and Pliquett 2007).

3.1   Electrical circuit

The circuit itself, as well as the real and imaginary parts of its impedance, is shown in Fig. 5. It can be seen that the measurement results of the IMPSPEC system do not differ significantly from those from a much more expensive analyzer from Agilent (HP4194A).

3.2   Dielectric constant of liquids

For this measurement, a coaxial probe was used as a sensor. The high-frequency M-sequence module was used in reflection mode (this is possible after introducing a directional bridge into the original system). Accordingly, the output of the properly calibrated UWB AU is the reflection coefficient of the material under test. From this data, the complex impedance and dielectric constant of liquids can be easily calculated:

$$\underline{Z}(f) = Z_0 \frac{1 + \underline{R}(f)}{1 - \underline{R}(f)} \tag{1}$$

$$\underline{\varepsilon}'(f) = \frac{1}{C_0 j \pi f \underline{Z}(f)} \tag{2}$$

In (1) and (2) Z_0 is the impedance of coaxial probe ($Z_0 = 50\ \Omega$) and C_0 is its capacitance ($C_0 = 2 \cdot 10^{-14}$ F).

Real parts of the dielectric constant versus frequency for a number of liquids calculated from measured data as described are shown in Fig. 6. Evidently, results of IMPSPEC do not

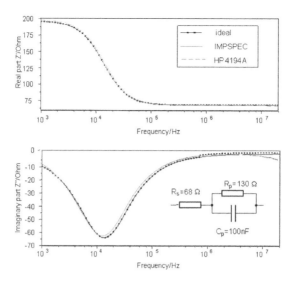

Figure 5.   Real and imaginary parts of impedance, measured and calculated.

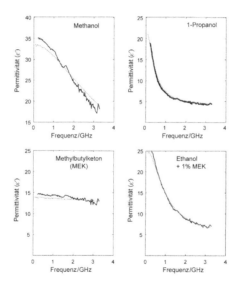

Figure 6.   Dielectrical permittivitz of liquids, real part, measured with vector network analzyer (red) and UWB AU (black).

deviate from the results measured with the laboratory network analyzer more then a few percent and both devices give results, similar to those described in literature (Kaatze 2007).

3.3   *Contents of carbon nanotubes in polymer material*

The sketch of measurement scenario and results are shown in Fig. 3 and Fig. 7. A horn antenna, operating in the frequency range of about 1–4.2 GHz was directed at different pieces of opaque polymer material with different content of carbon nanotubes. For clarity of representation, the impulse response of the material with lowest nanotube content was subtracted from all other datasets. Evidently, resulting spectrums of measured impulse responses depend on the nanotube contents strongly and thus, they can be used to identify this quantity.

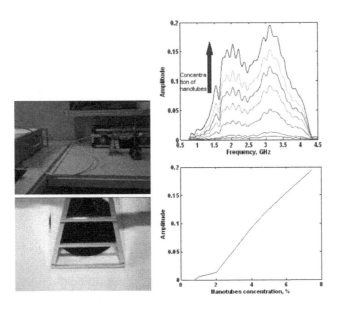

Figure 7. Measurement system in use for determining the content of carbon nanotubes in polymer materials: photos and results.

The simplest way is to use for this purpose empirical dependency between measured signal amplitudes and actual nanotube content (Fig. 7).

4 CONCLUSION

The modular concept of broadband impedance spectroscopy analyzers is presented. Configuration of hardware and software can be relatively easily adjusted to meet a broad range of particular requirements for the specific measurement task. Therefore, the system can be applied in process monitoring to solve a number of problems. The system has high measurement speed, good long-term stability and is relatively easy to use. This opens up new applications (such as in moisture and liquid sensing, non-destructive testing of polymers, etc.) which are difficult to implement with state of the art measurement techniques.

REFERENCES

Kaatze, U. (2007). Reference liquids for the calibration of dielectric sensors and measurement instruments. *Measurement Science and Technology 18*, 967–976.

Nacke, T., A. Barthel, J. Friedrich, M. Helbig, J. Sachs, P. Peyerl, and U. Pliquett (2007). A new hard and software concept for impedance spectroscopy analyzers for broadband process measurements. In *ICEBI 2007, IFMBE Proceedings 17*.

Sachs, j., P. Peyerl, and R. Zetik (2003). Stimulation of UWB-Sensors: Pulse or Maximum Sequence? In *International Workshop on UWB System*.

Corrosion

Lecture Notes on Impedance Spectroscopy – Kanoun (ed)
© *2011 Taylor & Francis Group, London, ISBN 978-0-415-68405-7*

The impact of decreasing chloride concentration on the corrosion state of steel in alkaline solution

T. Eichler, A. Faulhaber, K. Weidauer & B. Isecke
Federal Institute of Materials Research and Testing (BAM),
Division VI.1 "Corrosion and Corrosion Protection", Germany

ABSTRACT: The initiation of pitting corrosion on unalloyed steel in alkaline solutions depends unchallengeable on the $c(Cl^-)/c(OH^-)$-ratio. Numerous investigations on the critical chloride concentration for pit initiation can be found in literature (Böhni 1974; Breit 1998; Breit 2003; Glass and Buenfeld 1997; Sagüés and Li 2001; Page 2009). Investigations on the repassivation and the role of chloride concentration during repassivation of mild steel are rare (Sagüés and Li 2001; Eichler, Isecke, and Bäßler 2009). During cathodic protection (CP) of reinforced concrete structures the necessarily applied electric fields causes the migration of chlorides (Andrade, Sanjuán, Recuero, and Rio 1994; Buenfeld, Glass, Hassanein, and Zhang 1998; Luping and Gulikers 2007; McGrath and Hooton 1996) towards the anode, when the applied field is strong enough to overcompensate the back diffusion of chloride ions. Both, the migration of chlorides and the increase of pH at the steel surface due to forcing the cathodic reaction are well acknowledged to have beneficial impact on the efficiency of CP, (Bertolini, Bolzoni, Cigada, Pastore, and Pedeferri 1993; Bertolini, Pedeferri, Redaelli, and Pastore 2003; Glass and Hassanein 2003; Glass, Hassanein, and Beuenfeld 2001; P. Novák, Kouril, Msallamová, and Krticka 2007). The studies presented in this paper aim to clarify some open questions on role of chloride depletion at the steel surface in cathodic protection of steel in concrete. For this purpose steel samples in artificial concrete pore (ACP) solution while decreasing the chloride concentration were studied using galvanotstatic polarisation tests and electrochemical impedance spectroscopy (EIS). The evaluation of impedance spectra by equivalent circuit fitting gives evidence that the effect of decreasing the chloride concentration of the test solution rather results in sustainable reduction of corrosion rates than in regaining the state of passivity.

Keywords: cathodic protection, corrosion, repassivation, decreasing chloride concentration, steel, alkaline solution, secondary protection system, macro cell

1   INTRODUCTION

Usually, the reinforcement of reinforced concrete structures is well protected against corrosion by a dense and pore free oxide layer, the so called passive layer that forms on the steel surface in the highly alkaline environment, which is provided by the surrounding concrete. This layer has only a thickness of a few nanometres but it almost completely suppresses the anodic dissolution of iron (1).

When chlorides from external sources like de-icing salts or seawater etc. take ingress into the concrete, the passive layer may become instable if a critical chloride concentration is exceeded at the steel surface. The result of this process is local depassivation followed by pitting corrosion. The depassivated reinforcement acts as anode, where the dissolution of iron (1) takes place, while the remaining passive steel surfaces form cathodes (2).

$$Fe \rightarrow Fe^{2+} + 2e^- \tag{1}$$

$$O_2 + 2H_2O + 4e^- \rightarrow 4OH^- \tag{2}$$

In following reaction-steps corrosion products arise at the anode (4) and (5).

$$Fe^{2+} + 2Cl^- + H_2O \rightarrow Fe(OH)Cl + HCl$$
$$4Fe(OH)Cl + 2H_2O + O_2 \rightarrow 4FeO(OH) + 4HCl \tag{3}$$

$$Fe^{2+} + 2OH^- \rightarrow Fe(OH)_2$$
$$4Fe(OH)_2 + O_2 \rightarrow 4FeO(OH) + 2H_2O \tag{4}$$

The resulting potential difference between the anode and cathode provides the driving force for macro cell corrosion. The macro cell current is the dominating part of the total corrosion current (Raupach 1992) which can be expressed by (5).

$$I_{corr} = \frac{E_c - E_a}{\sum R_\Omega} + I_{self} \tag{5}$$

In order to control corrosion in technical systems, as for example in the case of chloride induced pitting corrosion of the reinforcement in reinforced concrete structures and to reduce remaining corrosion rates to negligible values, since more than 30 years cathodic protection is a well established and approved rehabilitation measure. The reinforcement of the building protected by CP is subjected to a cathodic current that is introduced by an external retrofitted anode. The previously existing potential differences between anodic and cathodic surface areas are reduced by the cathodic polarisation which is caused by the cathodic current flow from external anode to the reinforcing steel. Considering the characteristic polarization curves for active and passive iron in aqueous solutions, schematically shown in Fig. 1, it becomes obvious that a comparable small potential shift to cathodic direction can reduce corrosion current densities of a corroding system to values, which are in the same order of magnitude as corrosion current densities of passive systems, which is, according to (Boukamp 1995) below 0.1 $\mu A/cm^2$.

In addition to the cathodic polarisation the so called secondary protection mechanisms occur, such as the increase of pH at the steel/concrete interface due to forcing the cathodic reaction (2) as well as the migration of chloride ions, caused by the necessarily applied electric field. A derived and modified Fick's law can be used to describe the flux of ions, when assuming the flux of ions to be the sum of diffusion and migration and in the case of one-dimensional penetration (6).

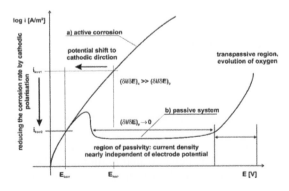

Figure 1.   Schematic representation of polarisation curves for active and passive iron in aqueous solution.

$$\vec{J} = -\vec{D}\left(\frac{\partial c}{\partial x} - \frac{zFU}{RTL} \cdot c\right) \qquad (6)$$

Where $\vec{J}$, is the total flux of ions (kg/m^2 s); $\vec{D}$ the diffusion coefficient (m^2/s); c the ion concentration (kg/m^3); x the distance from the surface exposed to the source solution of the ions (m); z the absolute value of ion valence, for chloride $z = 1$; F the Faraday constant, 9.648×10^4 J/V mol; U the absolute value of potential difference (V), cp. $E_c - E_a$ (5); R the gas constant, 8.314 J/K mol; T the temperature (K); L the thickness of specimen (m) and t is the time (s).

It is obvious that the electric field can effect a reduction in the chloride concentration at the steel surface and thus lead to a decrease of the important $c(Cl^-)/c(OH^-)$-ratio. These effects are well acknowledged to have a major impact on the effectiveness of cathodic protection of steel in concrete, although detailed research work into the various mechanisms and their effects on the corrosion state of technical systems is still missing.

For this reason, the investigations presented and discussed in the following aim to clarify the impact of decreasing the $c(Cl^-)/c(OH^-)$-ratio at the steel/environment interface on the corrosion state of mild steel samples in ACP-solutions.

2 EXPERIMENTAL SETUP AND RESULTS

2.1 *Experimental setup*

The presented test series were performed using artificial concrete pore (ACP) solution with pH 12.6, consisting of a saturated calcium hydroxide (Ca(OH)$_2$) solution with precipitates and additions of sodium hydroxide (NaOH) and potassium hydroxide (KOH) in a ratio of 1:6. The solutions had four different chloride concentrations in the range of 50 mmol/l and 1.250 mmol/l. The steel samples were made of BSt500 KR, which is reinforcing steel where the mill scale is mainly removed due to cold forming. The samples had a total surface area of 4.1 cm^2 and were electrical connected by spot welding a stainless steel (1.4370) welding wire with a diameter of 2.0 mm. The welding wire was coated by a lacquer in order to minimize its electrolytic contact to the test solutions. This specimen setup has proved to be expedient in order to prevent crevice corrosion at connecting point, cp. [Beck und Co.]. The galvanic coupling of stainless steel and unalloyed steel has no significant effect on the corrosion behaviour of the unalloyed sample, cp. [Abreu]. Table 1 gives an overview of the different test solutions and the appendant chloride concentration at the beginning of the test. Each test series consists of eight nominal equal specimens (series 126000 three samples were used as passive reference samples), which were polarised anodically with constant current density of 36.6 µA cm^2. Anodic polarisation was required in order to enable recording the changes in corrosion state during polarisation in consequence of decreasing the chloride concentration. For this purpose up to eight Bank MP 81 potentiostats in the galvanotstatic mode were used. During the polarisation time currents and potentials were permanently recorded using an AGILENT 34970A Data Acquisition/Switch unit. All potential values are related to the saturated Ag/Ag(Cl$^-$) reference electrodes. The samples were pre-corroded for 72 h at the initial chloride concentration. In each series, with exception of 126000, at five of eight samples the test solutions were diluted

Table 1. Survey of the test solutions used and size of the test series.

Test series	Number of samples	pH	$c(Cl^-)_{init}$ (mmol/l)	$c(Cl^-)/c(OH^-)$ -ratio
126000	3	12.6	0	–
126050	8	12.6	50	1.26
126100	8	12.6	100	2.51
126150	8	12.6	150	3.77
1261250	8	12.6	1250	31.4

every 72 h by substitution of 300 ml of the total volume of 800 ml through chloride free AGP solution with pH 12.6. This corresponds to a concentration reduction factor of $\zeta = 0.625$. In test series 1261250 a reduction factor of $\zeta = 0.375$ was used. At every concentration step the chloride concentration was determined by Ion Chromatography.

2.2 Results

Figure 2 shows exemplarily the records of the electrode potential development of the test series 1261250 (samples 1–5 at gradually decreasing chloride concentrations and samples 6–8 at constant chloride concentration). The first vertical red line on the left side marks the beginning of the experiment; each remaining line represents one dilution step. Initially, the samples show active behavior at a potential of about −500 mV. Considering to the entire potential record a transition phase between active and "passive" behaviour can be observed between the fourth and fifth dilution step. These thresholds, −500 mV for active behavior and 500 mV for passive behavior and the development of electrode potentials were also found in the remaining test series.

Figure 3 shows exemplarily the graphical evaluation of Fig. 2, the development of the mean polarisation potentials over time together with the chloride concentration of the test solution within the shown time intervals for test series 1261250.

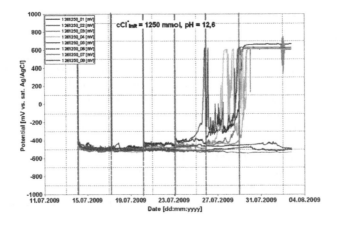

Figure 2. Development of electrode potential over the time against chloride concentration of the test series 1261250.

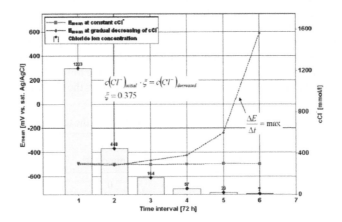

Figure 3. Development of electrode potential (mean value) over the time against chloride concentration of the test series 1261250.

The time interval with the maximum slope of the potential curve was defined as "repassivation" interval, cp. Fig. 3. An overview of the final chloride concentrations of the all test series is shown in Table 2.

Subsequent to the anodic polarisation tests and after depolarisation the samples were examined by electrochemical impedance spectroscopy (EIS) at open circuit potential (E_{oc}) in order to get more detailed information on the corrosion state. The impedance spectra were recorded using a GAMRY Reference 600 system. The evaluation of the measured impedance data was carried out by equivalent circuit fitting using the EchemAnalyst Software. All impedance spectra were reviewed regarding the steady state of the system by Kramers-Kronig-transformation (?), which is implemented in the EchemAnalyst software. All impedance spectra were recorded in a frequency range between 10^5–10^{-2} Hz, the applied DCvoltage was 0 V vs. E_{oc} and the AC amplitude was 10 mV$_{RMS}$.

Figure 4 shows the appropriate equivalent models, which were used for fitting the measured impedance spectra using the EchemAnalyst Software. The circuits used for fitting as well as the fitting procedure are reported elsewhere (Eichler, Isecke, and Bäßler 2009; Feliu, Gonzáles, Adrade, and Feliu 1998b; Feliu, Gonzáles, Adrade, and Feliu 1998a; Sánches, Gregori, Alonso, Garcia-Jareño, Takenouti, and Vicente 2007; Sánches, Gregori, Alonso, Garcia-Jareño, and Vicente 2006).

In the recent literature numerous more or less different, equivalent circuits with various numbers of time constants can be found. The challenge is to find a suitable equivalent circuit where the time constants have reasonable physical descriptions.

The evaluation of the impedance data of the chloride containing series was performed using three different equivalent circuits, cp. Fig. 4b)–4d). Expectedly, circuit 3 with two time constants fits in most cases the impedance spectra of series 126150 and 1261250 better than circuit 2 with only one time constant.

Table 2. Overview about the determined repassivation chloride concentration in the chloride intended test series.

Test series	$c(Cl^-)_{init}$ (mmol/l)	ζ	Number of dissolutions till repassivation concentration	$c(Cl^-)_{rep}$ (mmol/l)
126050	50	0.625	3	21–13
126100	100	0.625	4	25–16
126150	150	0.625	5	24–15
1261250	1250	0.375	5	20–7

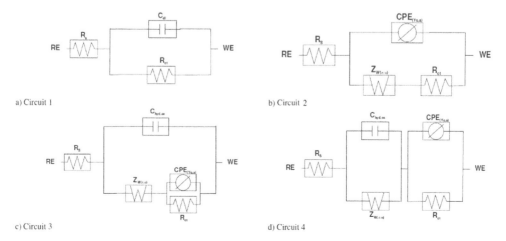

a) Circuit 1   b) Circuit 2

c) Circuit 3   d) Circuit 4

Figure 4. Used equivalent fitting models for the test series 126000–1261250.

The presented results in Fig. 5 show the Bode-plots of the passive test series (126000), which are used as reference for the chloride containing test series. The kink at 50 Hz in the phase angle curve was caused by the mains frequency. Figures 7 to 10 show exemplarily one sample from each series (active and "re-passivated"). For the purpose of comparison one sample of series 126000 was added to each Bode-plot of the remaining test series.

The high values of the phase angle in the low frequency range ($<10^{-1}$ Hz) emphasize the strong capacitive behaviour of the system and are strongly indicating passive conditions. In this case fitting was carried out using circuit 1, cp. Fig. 4, which delivered the following fitting parameter: $R_{ct,mean} = 337.6$ K cm^2, $C_{dl,mean} = 4.00 \cdot 10^{-4}$ F/cm^2 and $E_{OC,mean} = -2$ mV vs. Ag/AgCl, cp. Table 3, while no significance of diffusion could be found. The R_s of all test solutions are in the expected region of $R_{s,mean} = 1$–22 Ω.

Considering the results presented in Table 3 it can be concluded that the samples of the chloride reduced solutions (sample 1–5) have higher charge transfer resistance than the samples measured in undiluted solutions (active samples 6–8). Nevertheless, the charge transfer resistance of the "re-passivated" samples is still small compared to the passive samples.

According to (7), the combination of increasing Rct- and s-values results in increasing polarisation resistances and thus reducing significantly the remaining corrosion rate, cp. (5).

$$R_p = \sum_{i=1}^{n} R_i = R_s + Z_{diff,\omega \to 0} + R_{ct} \qquad (7)$$

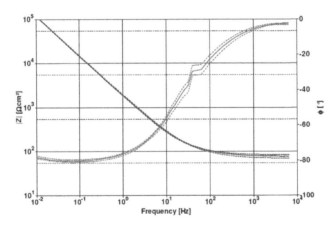

Figure 5.   Bode plots of the three samples from the test series 126000.

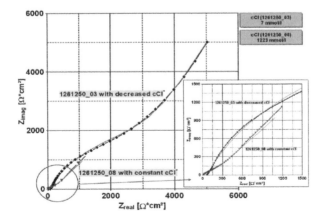

Figure 6.   Nyquist Plots; 1261250; active and re-passivated samples.

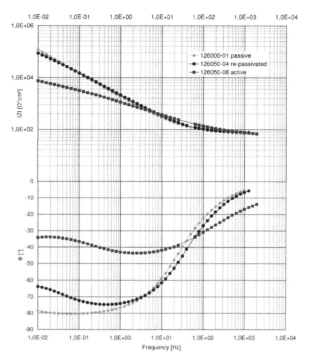

Figure 7.   Bode plots, exemplarily samples of a passive (series 12600), re-passivated and active reinforcement steel sample from series 126050.

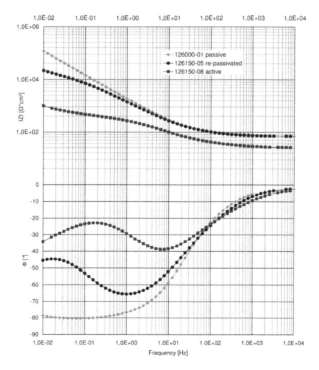

Figure 8.   Bode plots, exemplarily samples of a passive (series 126000), re-passivated and active reinforcement steel samples from series 126100.

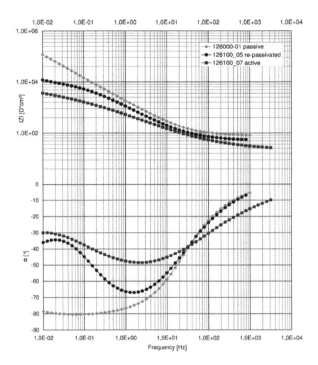

Figure 9. Bode plots, exemplarily samples of a passive (series 12600), re-passivated and active reinforcement steel samples from series 126150.

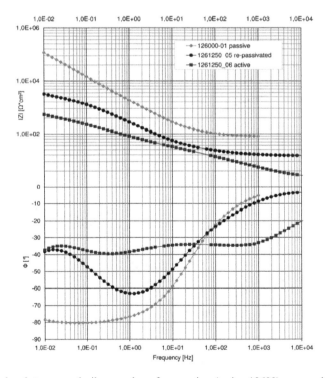

Figure 10. Bode plots, exemplarily samples of a passive (series 12600), re-passivated and active reinforcement steel samples from series 1261250.

Table 3. Mean values of the system parameters, determined by equivalent circuit fitting of impedance data.

Time series including sample number	$R_{s,mean}$ (Ω)	$Y_{0,mean}$ (S · sa/cm²)	a_{mean}	σ_{mean} $\left(\Omega \cdot cm^2 \sqrt{s}\right)$	$ct, mean$ (kΩ·cm²)	$E_{OC,mean}$ (mV)
126050_01-05	19	1.5E-4	0.819	1.15E+4	129.8	−171
126050_06-08	11	2.2E-4	0.615	1.86E+3	4.7	−430
126100_01-05	14	2.0E-4	0.799	2.92E+3	7.7	−237
126100_06-08	5	4.27E-4	0.642	6.64E+2	2.9	−429
126150_01-05	17	2.07E-4	0.749	3.61E+3	9.2	−169
126150_06-08	7	1.60E-3	0.661	1.66E+2	0.3	−539
1261250_01-05	19	2.49E-4	0.743	2.67E+3	7.1	−257
1261250_06-08	1	1.00E-3	0.437	4.61E+3	0.8	−529
126000_01-03	22	$C_{DL,mean} = 4.00$E-4 F/cm²			337.6	−2

3 CONCLUSION AND OUTLOOK

The anodic polarisation tests have shown that the severity of pitting corrosion on reinforcing steel can be lowered when reducing the chloride concentration of the test solution. For all tested conditions the c(Cl⁻)$_{crit}$ for $\Delta E/\Delta t$ = max is in the same order of magnitude and in the range of 7 and 25 mmol/1.

On the basis of the impedance spectra and the equivalent circuit modelling it could be figured out that although no repassivation in a commonly used context could be reached by decreasing the chloride concentration, a sustainable reduction of the remaining corrosion rate of the system could be obtained.

Future research work will have to show, if the results found by anodic polarisation tests are also valid for cathodic olarisation tests. The impact of increasing the pH at the steel/medium interface is subject of another test series in the framework of the presented research project.

The cathodic partial process with respect to the ratio of reduction of oxides and oxygen has to be investigated in order to enable the implementation of such data in the mathematical model.

The transferability to practical systems and to steel in concrete respectively will have to be demonstrated by adequate test series.

ACKNOWLEDGEMENTS

This work is a first part of the joint research project "Numerical Simulation of Cathodic Protection of the Rear Reinforcement of Reinforced Concrete" which is funded by the DFG (German Research Foundation). The Authors are grateful that the DFG is supporting this research work.

REFERENCES

Andrade, C., M.A. Sanjuán, A. Recuero, and O. Rio (1994). Calculation of chloride diffusivity in concrete from migration experiments, in non steady-state conditions. *Cement and Concrete Research 24*, 1214–1228.

Bertolini, L., F. Bolzoni, A. Cigada, T. Pastore, and P. Pedeferri (1993). Cathodic protection of new and old reinforced concrete structures. *Corrosion Science 35*, 1633–1639.

Bertolini, L., p. Pedeferri, E. Redaelli, and T. Pastore (2003). Repassivation of steel in carbonated concrete induced by cathodic protection. *Materials and Corrosion/Werkstoffe und Korrosion 54*, 163–175.

Böhni, H. (1974). Die Bestimmung von Lochfrapotentialen unter besonderer Bercksichtigung der galvanokinetischen Memethodik. *Materials and Corrosion/Werkstoffe und Korrosion 25*, 97–104.

Boukamp, B.A. (1995). A Linear Kronig-Kramers Transform Test for Immittance Data Validation. *Journal of the Electrochemical Society 142*, 1885–1894.

Breit, W. (1998). Kritischer Chloridgehalt Untersuchungen an Stahl in chloridhaltigen alkalischen Lösungen. *Materials and Corrosion/Werkstoffe und Korrosion 49*, 539–550.

Breit, W. (2003). Kritischer korrosionsauslösender Chloridgehalt Untersuchungen an Mörtelelektroden in chloridhaltigen alkalischen Lösungen. *Materials and Corrosion/Werkstoffe und Korrosion 54*, 430–439.

Buenfeld, N.R., G.K. Glass, A.M. Hassanein, and J.-Z. Zhang (1998). Chloride transport in concrete subjected to electric field. *Journal of Materials in Civil Engineering 10*, 220–228.

Eichler, T., B. Isecke, and Bäßler (2009). Investigations on the re-passivation of carbon steel in chloride containing concrete in consequence of cathodic polarisation. *Materials and Corrosion/Werkstoffe und Korrosion 60*, 119–129.

Feliu, V., J.A. Gonzáles, C. Adrade, and S. Feliu (1998a). Equivalent circuit for modelling the steel-concrete interface. I. experimental evidence and theoretical predictions. *Corrosion Science 40*, 975–993.

Feliu, V., J.A. Gonzáles, C. Adrade, and S. Feliu (1998b). Equivalent circuit for modelling the steel-concrete interface. II. Complications in applying the stern-geary equation to corrosion rate determinations. *Corrosion Science 40*, 995–1006.

Glass, G.K. and N.R. Buenfeld (1997). The presentation of the chloride threshold level for corrosion of steel in concrete. *Corrosion Sience 39*, 1001–1013.

Glass, G.K. and A.M. Hassanein (2003). Surprisingly Effective Cathodic Protection. *The Journal of Corrosion Science and Engineering 4*, 7.

Glass, G.K., A.M. Hassanein, and N.R. Beuenfeld (2001). Cathodic protection afforded by an intermittent current applied to reinforced concrete. *Corrosion Science 43*, 1111–1131.

Luping, T. and J. Gulikers (2007). On the mathematics of time-dependent apparent chloride diffusion coefficient in concrete. *Cement and Concrete Research 37*, 589–595.

McGrath, P.F. and R.D. Hooton (1996). Influence of voltage on chloride diffusion coefficients from chloride migration tests. *Cement and Concrete Research 26*, 1239–1244.

Novák, P., M. Kouril, S. Msallamová, and S. Krticka (2007). Cathodic Passivation. In *1st International Conference "Corrosion and Material Protection"*.

Page, C.L. (2009). Initiation of chloride-induced corrosion of steel in concrete: role of the interfacial zone. *Materials and Corrosion/Werkstoffe und Korrosion 60*, 586–592.

Raupach, M. (1992). *Zur chloridinduzierten Makroelementkorrosion von Stahl in Beton*, Volume 433 of *Deutscher Ausschuss fr Stahlbeton*. Beuth Verlag GmbH.

Sagüés, A.A. and L. Li (2001). Metallurgical effects on chloride ion corrosion threshold of steel in concrete. Technical report, Department of Civil and Environmental Engineering, University of South Florida.

Sánches, M., J. Gregori, M.C. Alonso, J.J. Garcia-Jareño, H. Takenouti, and F. Vicente (2007). Electrochemical impedance spectroscopy for studying passive layers on steel rebars immersed in alkaline solutions simulating concrete pores. *Electrochimica Acta 27*, 7634–7641.

Sánches, M., J. Gregori, M.C. Alonso, J.J. Garcia-Jareño, and F. Vicente (2006). Anodic growth of passive layers on steel rebars in an alkaline medium simulating the concrete pores. *Electrochimica Acta 52*, 47–53.

Lecture Notes on Impedance Spectroscopy – Kanoun (ed)
© 2011 Taylor & Francis Group, London, ISBN 978-0-415-68405-7

Author index

Printed and bound by CPI Group (UK) Ltd, Croydon, CR0 4YY

18/10/2024

01776252-0005